Photography in Printmaking

Charles Newton

V&A / Compton / Pitman

ISBN: 273 01470 6

Published for the Victoria and Albert Museum Photography in
Printmaking exhibition
by The Compton Press Ltd, The Old Brewery,
Tisbury, Wiltshire
and
Pitman Publishing Ltd,
39 Parker St., London WC2 5PB.

The exhibition will also be shown at the Mappin Art Gallery, Sheffield
and the Laing Art Gallery, Newcastle-upon-Tyne

Front cover: Allen Jones, *No. 2 from Life Class*, 1968, lithograph
and photo-lithograph

Back cover: Nils Burwitz, *Namibia, Heads or Tails*,
1979, screenprint

Designed by Humphrey Stone
Printed by The Hillingdon Press in England

Contents

Foreword

A few years ago it became apparent in the print shows of art schools and West End galleries that photography had assumed a major role in contemporary printmaking. Either photographic techniques or photographic images or both were used in printmaking with increasing frequency and in abundant variety. It seemed to us worth while to seek the reasons and to analyse the results of this marriage of photography and printmaking. Clearly it was essential to trace the relationship back to the beginnings of photography in the early 19th century, but in order to keep the subject within the bounds of a single exhibition it was felt best to treat this part of the story in outline only and to place the emphasis on the 20th century, particularly on the period since 1945.

It is a subject never previously investigated, and credit must go in the first place to the exhibition organisers, Susan Lambert and Charles Newton for threading their way through a mass of material of great technical and aesthetic complexity to present a coherent and comprehensible show. They interviewed artists to investigate the varying techniques and different purposes for which photography was applied and thanks must be expressed in particular to Chris Betambeau, Nils Burwitz, Bob Chaplin, Richard Hamilton, Stanley Jones, Eduardo Paolozzi, Chris Prater, Michael Rothenstein and Cliff White, the print makers and printers concerned, for giving so readily of their time.

It is a pleasure, also, to thank all the lenders who responded with characteristic generosity to our requests and particular mention must be made of the Science Museum from which a considerable proportion of the 19th century material has been drawn. For their help in the preparation of the exhibition and of the catalogue we are much indebted to Karen Amiel, Mark Haworth-Booth, Pat Gilmour, John Lee, Dorothy Newton, Sarah Postgate, Stella Ruff, Ray Smith, Joe Studholme, Sylvie Turner, John Ward and David Wright.

C. M. Kauffmann
Keeper of Prints, Drawings & Photographs,
and Paintings

Introduction

Prints are difficult to define. In his book *Prints and Visual Communication*
William Ivins defines prints, including photographs, as 'exactly repeatable
pictorial statements about anything . . .'.[1] This definition cannot be made more
specific without adding qualifications and exceptions. Nevertheless a
distinction can be drawn between a print made in a traditional way and a
photographic print as most people now understand it. To lessen the possibility
of confusion in what follows, a print which is made by one of the traditional
processes involving the printing of an image in ink on paper or some other
surface will be called a traditional print or just a print; a print made in this way
but involving the use of photography in the making of the printing surface will
be called a photo-mechanical print; while a photographic print, nowadays
usually thought of as a positive image formed by the darkening of silver salts
in a gelatine emulsion on paper, will always be referred to with its qualifying
adjective as a photographic print.[2]

Until the invention of photography, printmaking with blocks, plates and
stones worked by hand was the only means available of making exactly
repeatable visual images. This meant that prints were called upon to fulfil a
wide range of educational, religious and propagandist functions, many of
them mundane. Few prints were intended to be works of art. During the 19th
century the development of photo-mechanical printmaking processes
eventually caused the disappearance of the workaday role of the printmaker.
There is an analogy to this development in the results of the introduction of the
printing press to Europe in the 15th century. Printing grew from a process that
at first merely threatened the livelihood of scribes, to become the principal
way of spreading verbal ideas. It led to unforeseen social changes and
cultural effects, historians ranking it with gunpowder and the translation of the
Bible in its importance. It may seem an exaggeration to claim that
photography has had a similar effect on the dissemination of pictorial material
but it is significant that virtually all the printed imagery that surrounds us and
assails us now is produced by a photo-mechanical process and is, itself,
composed of photographs. The moving images thrust at us by film and
television are also direct results of the discovery of photography.

The development of photography has been interwoven with the more
traditional printmaking techniques since its invention. The distinction we now
draw between a photographic image made by developing a negative on to
light-sensitive emulsion and those processes which use the same negative
translated into a form that could be printed in ink was not widely made in the
19th century. For example, a book could be illustrated in one edition by
laboriously pasted-in photographs and in another edition by photogravures

from the same negatives. Because early photographs were liable to fade, it seemed natural to try to immortalise the image from nature by printing it in stable carbon ink rather than in the ephemeral tones of light-sensitive emulsions. Some of the inventions that did this, such as the Woodburytype, succeeded in looking at first glance indistinguishable from photographic prints made with silver salts. Conversely some artists, even when they wished expressly to illustrate the photograph as a work of art, produced images in techniques that hid their origin. In order to provide cheap line illustrations of his photographs, O. G. Rejlander made a drawing from a photograph of models, posed as for an anecdotal painting. From this he produced a relief printing block by electrotyping a photographically produced intaglio reproduction of the drawing. This could be set up with type in the press and looked when printed like a conventional wood-engraving, although photography had intervened at every stage.[3]

Photography and printmaking were linked at this stage because printmaking provided a quick, economical way of making multiple copies of a photographic image. Later it was photography that speeded up printmaking. This close relationship has been mutually advantageous, resulting in the continuing expansion of photo-mechanical printing techniques. Yet until relatively recently these techniques have been used only for commercial work, where they are valued for their speed, accuracy and adaptability rather than for any special qualities they give to an image. Artistic statements were made in both photographic prints and in traditional techniques but seldom in the hybrid photo-mechanical processes. Not until the 1960s did artists begin, to any extent, to use the photo-mechanical printmaking processes, and even then their work was rejected by official bodies.

The definition of an original print, issued by the United Kingdom National Committee of the International Association of Painters, Sculptors and Engravers in 1963, stressed that a print must be pulled from 'one or more plates, stones, screens, wood blocks, lino blocks, etc, predominantly executed by the hand of the artist . . .', while in 1965 the French National Committee of Engraving stated that original prints must exclude 'any and all mechanical or photo-mechanical processes'. In 1967 Paolozzi, goaded by an article titled 'Minting Prints', felt bound to defend his use of these processes: 'by employing engineering methods the iconography of the sculptor can be extended far beyond the normal range of the traditionally trained studio-bound artist and the high technical standards of industrial commercial processes, including screenprinting, can provide a complexity and range of possibility impossible by normal art-craft printing methods. To-day, a superior technology always out-dates older methods. This is irrefutable. Art now is not free from this historical situation.'[4] This attitude obtained wider support in 1968 when the Printmakers Council agreed that: 'The artist-printmaker has the inalienable right to decide himself on the methods he uses to make a print. These can be autographic, mechanical or photographic. He can use any processes in existence to-day or which could exist in the future.'[5]

Only recently has the distinction between photographic and traditional printmaking in works of art become blurred. It is now accepted that there is no hierarchy of techniques and that an artistic statement may be made with equal validity as a photograph, as a print using traditional or photo-mechanical

processes, or in a mixture of these techniques. The 1979 International Print
Biennale held at Bradford has included photographs for the first time.
Selected artists were invited to submit prints. Photographs were not excluded
in the brief and a number of artists chose to be represented by photographic
prints.

Photo-mechanical printmaking in the 19th century

After a slow start in the first half of the 19th century, the photo-mechanical
revolution gathered momentum; by 1890 the making of prints had been
irrevocably changed by the impact of photography. The first effective step in
this development was taken by a Frenchman, Joseph Nicéphore Niépce.
Since the beginning of the 19th century attempts to produce pictures by the
action of light on various substances had been made but no one had
succeeded in fixing the images that resulted. In 1827 Niépce made a
reproduction of an engraved portrait (Fig. 1) by a method that combined
photographic principles with a traditional printmaking technique. Thus
printmaking by photographic means can be said to have preceded the
invention of photography itself. Although this process, like electricity, was at

Fig. 1   J. N. Niépce, *The Cardinal
d'Amboise*, 1827. Etched plate,
inked up

This image was the first successful photo-
mechanical reproduction. It pre-dated the
photography of Daguerre and Fox Talbot.
Niépce's use of light-hardening bitumen as a
resist on an etching plate was the first
practical application of photographic principles to printmaking. He had been experi-
menting in this field since 1813 and had
succeeded in fixing a blurred image, consisting of a faint film of bitumen on pewter plate,
of his house, from nature, in 1826.

Fig. 2   W. H. Fox Talbot, *Lace*, 1842.
Calotype negative

Fox Talbot discovered a way of fixing the image made by the darkening of silver salts on exposure to light as early as 1843. He made photogenic drawings (his term) by placing objects similar to this piece of lace on light-sensitised paper and exposing them, creating a kind of shadow picture in negative. By 1840 he had discovered a way of developing a latent image, thus speeding up the time required for exposure and making possible the negative and positive process which he called Calotype or Talbotype.

first regarded as a scientific toy of not much practical use, it is the basis of all subsequent photographic process work in printmaking.

Niépce took a traditional print and waxed it to make the paper translucent. He applied the treated print to a metal plate which had been coated with bitumen of Judea, a tarry substance which hardens on exposure to light; he then exposed plate and print to bright light for a lengthy period. When the paper was removed from the plate, the bitumen had hardened where it had been exposed, while remaining soluble where it had been protected from the light by the lines of the print. These soft areas were dissolved away, leaving bare metal exposed in a network of lines reproducing the design of the original. The plate was then etched, and after removal of the remaining bitumen, inked and printed like a traditional etching. Thus, a reproduction of the original print was produced without any laborious hand-work.

Niépce's process required a printed image as a starting point. The first invention to compete with the illustrator printmaker's power to convey visual information was the Daguerreotype, which was made public in 1839. Using materials reminiscent of alchemy, this process fixed a positive image from nature by the action of the light of the sun through a lens on to a highly polished light-sensitised silver-plated copper sheet. The detail of the images so produced amazed the world but the process had its drawbacks. As the images were formed by minute changes in the polished surface of the silver they could only be seen when the plates were held at an angle to catch the light. Moreover, the plates were fragile and each Daguerreotype was unique: in order to make a replica the whole process had to be done afresh. These disadvantages were shortly overcome with the aid of traditional printmaking processes. In 1840 Dr Joseph Berres produced an ink-printed image from an etched solid silver Daguerreotype, and in the following year Hippolyte Fizeau

made electrotypes of Daguerreotypes, from which images could also be printed in ink.

The negative/positive photographic process, using the principle that silver salts darken on exposure to light, was discovered by W. H. Fox Talbot some years before he published it under the name of Calotype in 1841 (Fig. 2). The invention of this process meant that an image from nature could be produced entirely mechanically and repeated indefinitely without any intervention from a traditional printmaker. The value of photographic negatives and positives based on these principles was quickly appreciated by the printing trade, and descendants of the Calotype have become an essential ingredient in the majority of photo-mechanical printing processes.

The etched plate produced by Niépce inspired others to try to make photographically produced intaglio plates for the reproduction of works of art. Fox Talbot, in England, patented the use of bichromated gelatine instead of bitumen as a light sensitive resist in 1852. This idea was developed further in France under the name of heliogravure, known in England as photogravure. Used with an aquatint ground applied to the plate, it was able to reproduce tone. As a photograph was used to transfer the image to the plate, it did not, as in Niépce's method, destroy the original and could, therefore, be used to reproduce unique works such as paintings (Fig. 3). The technique was also used to make facsimiles of etchings and engravings of such quality that it is often difficult to distinguish the original from the photo-mechanically produced copy.

Another step towards mechanisation was the introduction of the cross-line screen during the 1880s. This invention was anticipated by Fox Talbot, who in 1853 reported that he had constructed a crude way of obtaining a tonal effect by putting two pieces of black gauze between the negative and the plate coated with light sensitive resist. He had also suggested the use of a glass screen etched with parallel lines. However this idea was not used commercially until a Czech, Karl Klič, perfected the idea, creating a screen of opaque squares and clear lines with which he printed the gelatine film before it was exposed under a transparent negative. Images printed in this way are entirely made up of dots. The dots are of equal size, but carry different thicknesses of ink, depending on how much light has passed through the negative at any point. If the screen is fine enough the dots are not easily visible, while their massing gives the effect of dark tones. The principle of the glass screen is used in lithography and relief printing as well.

At first photogravure plates were printed on flat-bed presses by hand. The time saved by preparing the plates by photography made the process viable enough for relatively expensive reproductions of popular paintings. Photogravure plates printed on a rotary press, using a screen where necessary, made it possible to produce high quality illustrations quickly, relatively cheaply, with enormous runs and photographic accuracy.

A parallel sequence of developments took place in the relief processes of printmaking. Wood-engraving was the first form of printmaking to use photography as an aid. Since this was the main way of illustrating books, periodicals and newspapers (because of the wood block's compatibility with type, both being relief processes) anything that saved time was an advantage. Not only were artists' drawings transferred photographically to the block, thus preserving the drawings without involving the time-consuming business of copying them by hand, but as early as 1839, the *Magazine of Science*[6] published some wood-engravings cut from photographic images thus removing one of the intermediary stages of interpretation in producing a pictorial record. However, the well tried conventions of the engraver of drawn images continued to dominate the style of the cutting.

Wood blocks were often electrotyped to enable the printing of longer runs than the wood itself allowed. This process – whereby copper was deposited electrochemically on a cast of a block so as to produce a metal relief block – was combined with photography in various ingenious ways in the 1860s. Experiments were made in forming relief blocks by electrotyping intaglio plates made of photographically-produced hardened bichromated gelatine. This elaborate process was finally superseded in the 1870s by the relatively simple line block process.

The line block is a printing block to which the design is transferred photographically in such a way that when the block is etched, the areas to be inked are left standing in relief. The speed with which it could reproduce a pen and ink sketch or a brush drawing made it popular with commercial artists. The line block could not produce tone, but this disadvantage was overcome in the 1880s by the application of the cross-line glass screen. The resulting half-tone letterpress block enabled any tonal image, photographed or drawn, to be printed on a wide range of papers including those of low quality. Illustrated newspapers had been available for a long time but the illustrations had been almost always reproduced by wood-engraving. It was not fortuitous that the rise of the popular press coincided with the availability on a massive scale of rapid mechanised methods of illustration.

Fig. 4   Henri Gaudier-Brzeska, *Christmas card: a monkey*, 1912. Line block

An elegant, spare and simple image exploiting the inherent characteristics of the medium.

By the end of the 19th century half-tone blocks were being printed in colour, using the principle of colour separation, making it possible to produce the books with brightly coloured illustrations which became so popular during the first decade of this century. By the turn of the century the range of available techniques for making inexpensive photo-mechanical prints was immense.

The planographic process of lithography had not long been developed before attempts were made to use photography in conjunction with it. Senefelder invented lithography in 1797 and Niépce experimented with a light-sensitive coating on the lithographic stone in the 1830s. But it was Alphonse Poitevin who discovered and patented in 1855 the process of applying albumen mixed with gelatine and potassium bichromate, the first really successful method of photo-lithography (Fig. 5). A photograph of an ink drawing could be reproduced by making a negative, placing it on a sensitised stone, exposing it to light and having removed it , washing away the soluble unhardened parts of the resist on the stone. The remaining coated areas acted like the greasy lithographic crayon, repelling water and picking up ink, so that the stone could be printed in the normal way. This technique did not become widely used until the invention of the half-tone screen.

Alphonse Poitevin patented another discovery in 1855. He found that a metal or glass plate coated with bichromated gelatine, exposed when dry under a negative and then moistened with water, would accept ink in proportion to the amount of light it received. This collotype, as it was called, had a printing surface of irregular crinkles formed as the gelatine dried in the initial platemaking process. When the moistened gelatine surface was inked and paper pressed on, a fine grained tonal image could be taken off. This process took some time to become established but under the names of albertype, autotype and later heliotype it became commercially viable in the 1870s. It gave unusually good quality tonal reproductions, even excelling the screened processes being developed at the same time.

The reaction in artistic circles

The overriding reaction to photography and photo-mechanical printmaking processes in most artistic circles was one of panic. The industry of reproductive engraving expanded at an enormous rate at just the moment when the new process which was to make its craft methods obsolete was being developed. Félix Bracquemond echoed the general dismay when he wrote: 'Photography has brought disruption to all that pertains to the art and craft of engraving. Disruption is hardly the word, considering this new wonder is doing away with engraving altogether'.[7] The livelihood of many of the more commercial printmakers was put in jeopardy. For example, the work of engraving maps from detailed drawings for the Ordnance Survey was taken over by photo-lithography and the reproduction of popular paintings could be more quickly and cheaply done by photogravure.

Illustrators did not fare much better. Those whose prime task was to convey pictorial information found that their drawn interpretations could not compete with the accuracy and relative objectivity of the photograph. Those who offered artistic vision in their illustrations discovered that they now had many more competitors. For artists who might not have been interested in wrestling with the problems of designing for reproduction in a technique that imposed its own limitations could now have their drawings, be they in line or tone, black or white, or in colour, reproduced with adequate faithfulness. Traditional printmaking techniques were now used only in special publications. Even wood-engraving became imbued with an artificial preciousness, and was used more by private presses than commercial publishers.

It was inevitable that the new techniques provided by photography would eventually encourage artists to adapt their work and their attitudes to it. With hindsight it seems however that the process took a long time. It is difficult to estimate the effect of photography on styles of illustration, but Beardsley in the 1890s appears to have been the first artist to make his name by deliberately exploiting the photo-mechanical processes for the specific qualities they give to a printed image. He made wash drawings especially for reproduction in the velvety tones of photogravure, and line drawings for the sharpness of the line block (Fig. 6). It is clear from his letters to his publisher that he was aware of the potential of these processes. In 1898 he wrote to Leonard Smithers: 'as you are giving plenty of reduction to the initials you must surely do the same by the other drawings or the pencil work will not look the same right away through the book. Reduction pulls a drawing together and strengthens it in every way.'[8] In another letter he said: 'the drawing is a trifle cold in the original, so please let the printer's ink on warm paper correct this fault.'[9] He knew the printing process could enhance the original. He saw his drawings as designs for the finished work, telling his publisher: 'Once blocks are made I don't worry any longer about possible loss of or accident to drawings.'[10] Using a system of dots, solid blacks and a crisp flowing line Beardsley broke away from the conventional cross hatching and shading that had persisted in the work of illustrators used to other media. He had many successors who in their turn exploited the increased range of effects possible with photo-mechanical printing processes.

Some fine artists did use photography for their own ends. It soon became commonplace for painters to work from photographs and there was a vogue

for composed and composite photographs that emulated paintings, anticipating the technique which became known as photomontage. The best known practitioners in the 19th century were the Swede, O. G. Rejlander and the Englishman, Henry Peach Robinson. These artists fitted together large photographic negatives of posed models in a kind of mosaic to imitate grand painterly scenes, achieving effects otherwise impossible in the early days of photography.

Another group including Millet, Corot and Rousseau scratched drawings on a piece of smoked glass.[11] The glass became translucent where the black pigment had been removed, creating a negative which could be printed directly on matt sensitised paper. As a simple, direct way of reproducing a drawing without elaborate equipment cliché verre or hyalography, as it was sometimes called, could not be bettered and it is surprising that it was not more widely practised (Fig.7). Perhaps its very simplicity and frank use of

Fig. 7   Camille Corot, *La Petite Soeur*.
Cliché verre

This print which has all the autographic qualities of Corot's work presents a conundrum for the art historian. Cliché verre fits neither into the category of a traditional print nor of a photographic process. Essentially it is a hand-drawn negative. It neatly demonstrates the current theory that the artist's idea and not the medium in which it is carried out is the work of art.

photography made it unacceptable to conventional artists. Only recently has it been taken up with some enthusiasm.[12]

Nevertheless, even among artists whose livelihoods were not directly affected by photography many had reservations about it. They were not altogether happy with the quality of photographically-based reproductions of their work. In 1888 Frith wrote: 'in photogravure, so popular now, the place of the second mind [i.e. the engraver] is supplied by a machine, which often does its best to destroy or mutilate the efforts of the first [i.e. the painter]; but sometimes, I admit, reproduces the effect of a picture with extraordinary accuracy. But accuracy, in my opinion, is not enough; I want at the same time the taste and skill, which amount almost to genius, with which the great engraver changes into black and white the colours of the picture before him.'[13]

They may well have felt threatened by the fashion for photographs which far exceeded that for traditional prints. It led Monsieur de Saint Arroman to make the pun 'The *cliché* is about to replace everything in art as well as in literature. Every day I come across people who look as if they were in their right minds, people like you and me, and these people candidly prefer a photograph to an engraving.'[14] To make matters worse photographers could not only produce pictorial records with greater speed, detail and accuracy than a traditional artist but they clearly thought of themselves as artists as well. Photographs were exhibited at the Paris Salon as an art form as early as 1855 and the photographic journals, like the *Illustrated Photographer* for which R. A. Seymour wrote a weekly advice column, were full of encouragement for amateurs to study pictorial composition in order to make better prints.

In the past, artists making prints primarily as artistic statements were in the minority. When the Royal Academy was founded in 1768, engraving was considered such a subordinate form of art that engravers were excluded altogether from its members.[15] Shortly afterwards, under pressure, six associate members were admitted but they were not allowed any voting powers nor were they allowed to teach engraving in the Academy schools. When in 1812 a committee was set up to investigate the inferior position held by engravers in the Academy, it reported that engraving was wholly devoid of 'those intellectual qualities of invention and composition, which painting, sculpture and architecture so eminently possess . . . its greatest praise consisting in translating with as little loss as possible those original arts of design . . .'. And so it concluded 'that with such an important difference in their intellectual pretensions as artists, it appeared to the framers of the Society that to admit engravers into the first class of their members would be incompatible with justice and due regard to the dignity of the Acedemy.'[16] This attitude held in official artistic circles gave later generations of printmakers, not admitted on equal terms with other members of the Academy until 1928, inherited psychological reasons for ensuring that their output should be seen to have nothing in common with facsimile work or commercial printing techniques.

Their work was influenced by a desire to establish printmaking as a form of art as original, creative and respectable as painting and sculpture. Seymour Haden's pamphlet *The relative claims of etching and engraving to rank as fine arts* issued in 1883 shows the ideas with which printmakers became preoccupied. In it he put forward the biased but prevailing opinion that etching (which among printmaking techniques had been the preferred medium of painters in the past), 'depending on brain impluse is personal . . . and has the various attributes which make up the sum of genius': it thus ranks as fine art, whereas engraving, tainted by its earlier use as primarily a reproductive medium, is 'without personality and all the attributes which attend the exercise of the creative faculty' and is, therefore, merely a craft.

Societies aiming to infuse new life into traditional printmaking were formed. In his preface to the first portfolio issued by one such society, the Society of Etchers set up in Paris in 1861, Théophile Gautier appealed to the public thus: 'in these times, when photography fascinates the vulgar by the mechanical fidelity of its reproductions, it is necessary to assert an artistic tendency in favour of free fancy and picturesque mood'. Artists were encouraged to draw on their own inspiration and imagination where, in the words of Gautier, the camera's 'brass-lidded eye of brass' could not follow, thereby stressing how different their prints were from photographs. In England the attitude was typified by Whistler who made a fetish out of the different effects achieved by inking the plate, denying the basic multiple nature of the print, and charged twice as much for individually signed impressions. The concept of the 'original' print came into being. Artists' prints began to be issued in Limited editions, sometimes with an artificial number of states, first creating the market for something rarified and precious, and then pandering to the collector who saw himself as possessing a particular kind of aesthetic sensitivity in his appreciation of them.

The twentieth century: mechanism and expression

In order to appreciate what photography has contributed to the immense changes that have taken place in printmaking in the 20th century one need only compare any 'fine art' etching of 1900 with Eduardo Paolozzi's

screenprint *Hollywood Wax Museum* (Plate B) of 1969. Both are printed in
ink on paper but they have little else in common. There is hardly a method or
concept that remains the same. How and why did printmaking change so
much?

After 1900 printmakers became split into two camps. Original printmakers,
like Sickert, represented the self-conscious craft tradition which depended on
a small market of connoisseurs. The other camp encompassed all forms of
printmaking that could not be defined as 'original' in the strictest sense: that is,
image-making methods that involved the use of technicians, automatic
machinery and photography. Although fine artists referred to these methods
under the opprobrious title of 'process' and were inclined to regard illustrators
with less respect if their work was reproduced by process, it was these
techniques that had the greater impact on the world at large. The amount of
visual information produced in the form of prints, available wherever a railway
or steamboat could deliver them, was unprecedented. For the first time,
millions in Kabul, Stoke Newington or the South China Sea could enjoy photo-
mechanically produced prints in newspapers and magazines. In one sense
Marshall MacLuhans's global village was already established. The change is
best appreciated by comparing a newspaper of 1804 with an illustrated
magazine or newspaper of 1904, the year in which the *Daily Mirror* first
appeared. While in the former simple columns of fine print give information in
a straightforward and literate way, in the latter images, most of them
ephemeral, jostle for position. They could include, for example, a reproduction
of an Academy painting next to an advertisement for processed food, a
photograph of the Russo–Japanese war, a sketch of the zoo, and cartoons
and photographs of politicians, all produced with elaborate photo-mechanical
care. Photo-process printmaking went some way to making it possible to
depict the rapid changes in the world, while itself altering the way in which
change was perceived. The modern reader's habit of rapidly skimming over
the surface of pictures rather than paying close attention to the words of a text,
came into being.

It was only as a by-product of the work of artists whose attitudes were a
direct response to the dramatic changes occurring in the world at large and
whose artistic output challenged accepted values in many media, that a link
was forged between commercial photo-mechanical printmaking and the craft
approach of 'original' printmakers. The idea of the ordered universe was being
challenged by philosophers, physicists and psychologists such as Russell,
Einstein and Freud, and the climate of thought was much influenced by the
advent of full-scale technological warfare. Ideas about reality broke
irrevocably away from their traditional forms. In painting, late 19th century
academic conventions seemed inadequate when an artist was confronted
with what was actually happening. Sentimental genre scenes and costume
heroics appeared irrelevant beside photographs of the hideous confusion in
the streets and on the battlefields of Europe. Yet photographs also had their
shortcomings. Pictures of horrors and trivialities were published under the
heading 'Events of the week' in an absurd patchwork. There was no hierarchy
of importance in the way images were presented and the equalising effect
was increased by the uniformity of the reflective surface of the page. Crashed
Zeppelins and tins of fruit salt shared a common reality. In this context the

pioneering of new ways of expression by the Cubist and early abstract artists and the forming of anti-art movements by the Futurists and Dada artists seems logical and right.

The new artists believed that the old media did not display the spirit of the age. The Futurists had tried to assimilate both machines and wars by seeing them as art in themselves. 'A racing car is more beautiful than the winged victory of Samothrace' and 'War is beautiful because it initiates the dreamt-of metalisation of the human body'[17] are typical quotations from Marinetti. Accepting the juggernaut of technological progress uncritically is no better than ignoring it and retreating into nostalgia for the past. But the Futurists' liking for machinery did help to secure recognition for machine-made pictures, a photo-mechanical image, in the days of tanks and aeroplanes.

Fig. 8   John Heartfield, *Hurrah die Butter is alle!* (Hurrah the butter is finished!) 1935. Photomontage

Heartfield worked in close collaboration with a photographer, setting up his own compositions with models as well as incorporating 'found' images. At exhibitions Heartfield insisted that the magazine in which the finished print appeared should be displayed alongside the photomontage. This print refers to Goering's infamous speech about 'guns before butter'.

Artists of the Dada movement – which according to the words of Hülsenbeck in the first manifesto of the Berlin group of 1918, called for an art 'which in its conscious content presents the thousandfold problems of the day, that art that has been visibly shattered by the explosions of last week, which is forever trying to collect its limbs after yesterday's crash . . .'[18] – were the first consciously and consistently to adopt photo-mechanical images in their work. They adopted them first in photomontage. Controversy surrounds the introduction of this art form. George Grosz claimed that he and John Heartfield (Fig. 8) invented it in 1916: 'On a piece of cardboard we pasted a mischmasch of advertisements for hernia belts, student song-books and dog food, labels from schnaps- and wine-bottles, and photographs from picture papers, cut up at will in such a way as to say in pictures what would have been banned by the censors if we had said it in words. In this way we made postcards supposed to have been sent home from the Front, or from home to the Front.'[19] On the other hand, Raoul Hausmann claimed to have invented it with Hannah Höch in 1918. These rival claims do not detract from Hausmann's statement in the article 'Definition de Foto-Montage' of 1932 that the Dadaists 'were the first to use photography to create, from often totally disparate spatial and material elements, a new unity in which was revealed a visually and conceptually new image of the chaos of war and revolution.'[20] The first dated photomontage appeared on the cover of the Dada periodical *Jedermann sein eigner Fussball* in February 1919. Designed by John Heartfield, it consisted of a composite of photographs showing a man with the body of a football. This was not only a photomontage but was reproduced by a half-tone relief block – a photo-mechanical process. The Dadaists adopted, as means of expression, the manifesto and the magazine. In using these means of spreading images, they inevitably found themselves making use of photo-mechanical processes.

The Surrealists too were constantly producing publications, illustrated by photo-mechanical means (Fig. 9), and like the Dadaists, they produced collages and photomontages as unique works. For example, Max Ernst composed collages from advertisements, technical manuals and photographs. The poet André Breton recalled the effect these had on his coterie: 'I remember very well the occasion when Tzara, Aragon, Soupault and I first discovered the collages of Max Ernst. We were in Picabia's house when they arrived from Cologne. They moved us in a way we never experienced again. The external object was dislodged from its usual setting. Its separate parts were liberated from their relationship as objects so that they could enter into totally new combinations with other elements.'[21]

The Dadaists and Surrealists were paralleled in Russia by the Constructivists, who aimed to use the elements of industrial technology to construct a new art for a new society. Both Rodchenko and El Lissitsky had come independently to the same conclusions about the value of photomontage. They also used it for propaganda and advertising, and simply for its modern, poetic and artistic appeal.

At the Bauhaus, Moholy-Nagy encouraged the publication of books with exciting images which epitomised the speed and headlong change of the new age. His own book, *Painting Photography Film* of 1925, described photomontage as 'an experimental method of simultaneous representation;

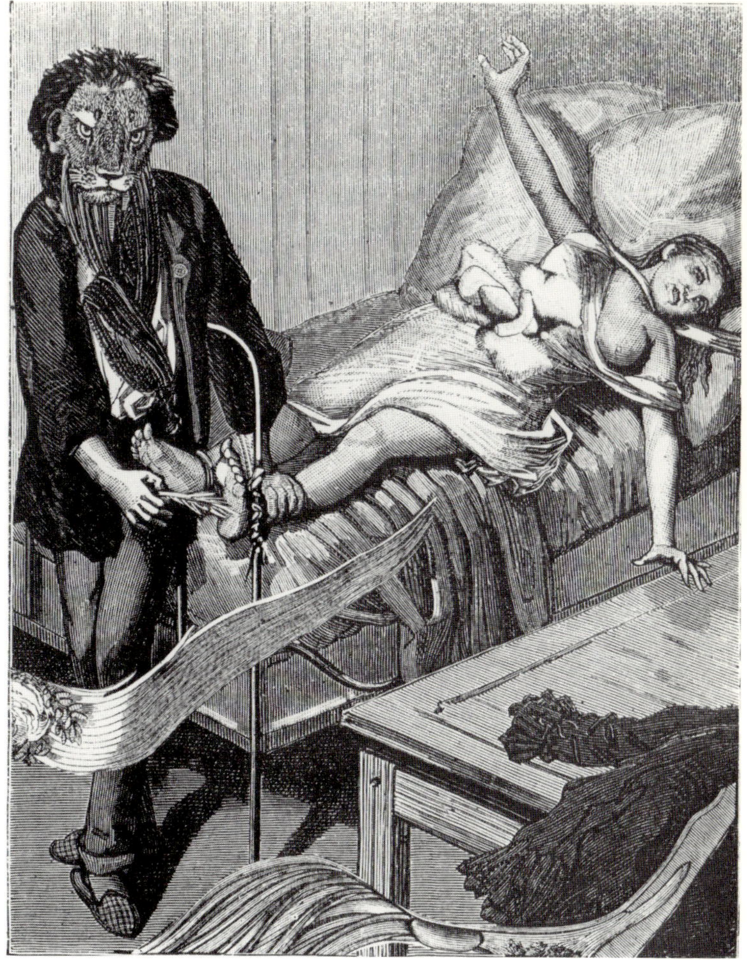

compressed interpenetration of visual and verbal wit; weird combinations of
the most realistic, imitative means which pass into imaginary spheres' and
stated most prophetically that 'it will soon be possible to do this work, at
present still in its infancy and done by hand, mechanically, with the aid of
projections and new printing processes'.

Photomontage was a radical rejection of the established canons of art and
yet, by the mid-1920s, it was widely used in certain avant garde circles. It was
not until forty years later, however, that photographic imagery and methods
were generally adopted in printmaking. This time gap requires some
explanation.

Photomontage was taken up with enthusiasm by advertisers (Fig. 10), who,
realising its propagandist power, ruthlessly pressed it into service to cater for
the demand for new imagery presented in new ways. Jan Tschichold's book
Die Neue Typographie of 1928 provided a compendium of noteworthy
examples of photomontage, including an advertisement for Pelikan ink by El
Lissitsky and reproductions of the work of Man Ray and Rodchenko. Another
book, *Photo-Eye* (Fig. 11), was recommended as a source of ideas in an

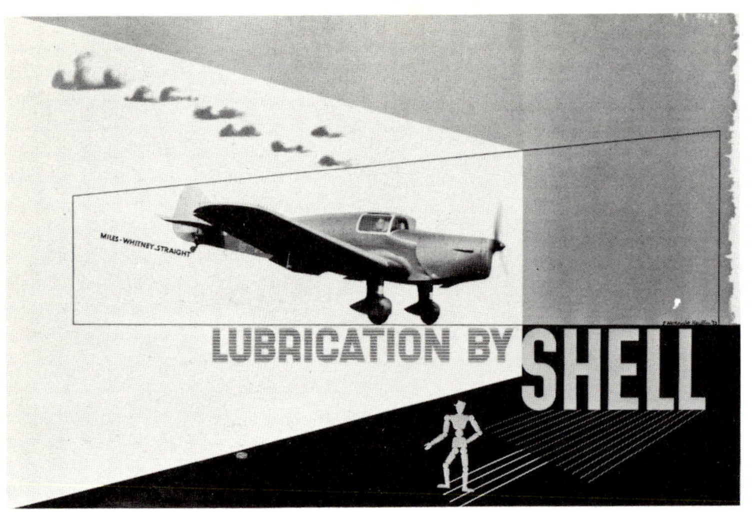

Fig. 10   E. McKnight Kauffer, *Poster advertising Miles-Whitney Straight Aircraft and Shell Lubricating Oil*, 1937. Offset lithograph

The image combines a photograph of an aircraft with painterly brushstrokes representing clouds. Cohesion is provided by the geometric outlines and the typography. The poster is a successful example of the use of photomontage in advertising, with the photographic detail of the plane contrasted with the minimal structure of the rest of the design.

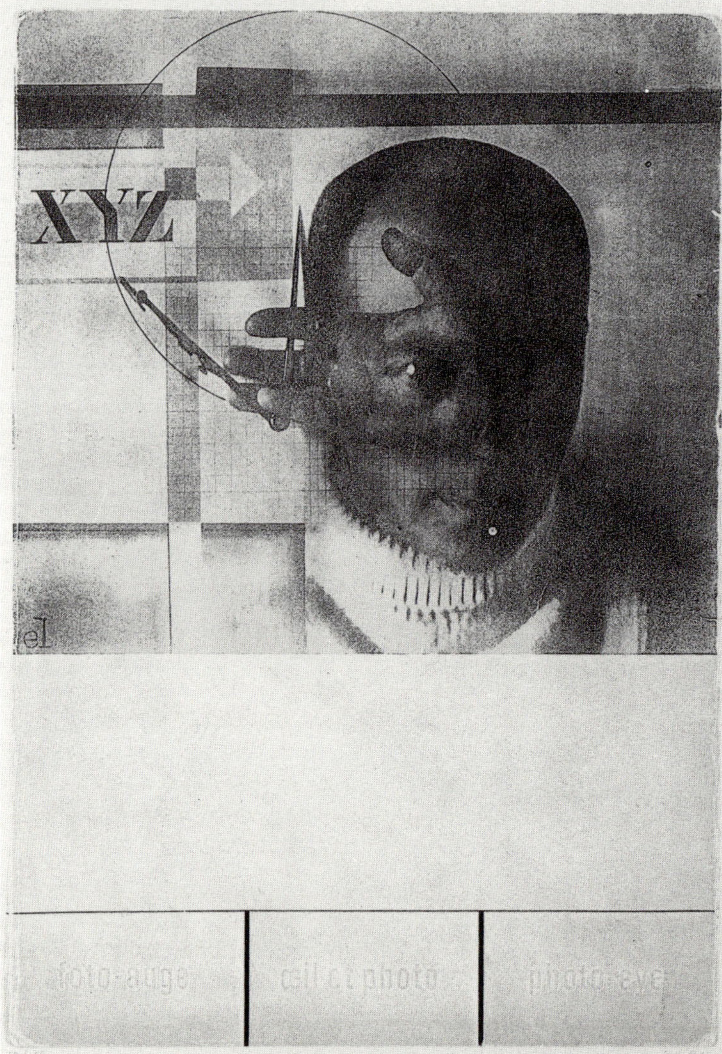

Fig. 11   R. B. Kitaj, *Photo-Eye*. From the series 'In Our Time', 1969. Screenprint

'In Our Time' is a series based on the covers of Kitaj's best loved books, which has the sub-title 'Covers for a small library after the life for the most part'. This print reproduces another print of major importance to the development of the use of photography in printmaking. The image is a self-portrait of El Lissitsky, 'The Constructor' of 1924, which was used as the cover illustration to the book *Photo-Eye* by Franz Roh and Jan Tschichold, published in 1929. This book reproduced photographs, photomontages and photograms, and became a source for the new generation of advertisers, designers and commercial artists who wished to use photography creatively.

Fig. 12   John Piper, *Chapel of St George, Kemptown*, 1939. Etching and aquatint

Piper was an exception, gaining respect for commercial printing through his involvement with the publication of *Axis*, a quarterly review of contemporary 'abstract' painting and sculpture. Blocks of type, used to suggest the brickwork of the church are, at the same time, redolent of the prayer-book.

article entitled 'What the camera can do for you' in *Commercial Art* in 1930.[22] Although this commercial application resulted in the development of improved colour printing, the association between photomontage and the sale of goods appears to have tainted it in the eyes of fine artists. The advertiser's aim, however it was dressed-up, was to encourage spending. Many artists agreed with William Blake: 'where there is any view of money art cannot exist, but war only . . .' In addition photomontage was unattractive to traditional artists because it did not show off manual dexterity to the same extent as drawing. Indeed the word itself had been adopted by the Dadaists in order to bring out this distinction. Hausmann said: '. . . it embodied our refusal to play the part of the artist. We regarded ourselves as engineers, and our work as construction: we assembled [in French: monter] our work like a fitter.'[23] The whole-hearted use of photographic imagery and photo-mechanical printing techniques in commercial work appears to have reinforced the determination of most original printmakers to make prints that were traditional in concept and execution (Fig. 12).

The print boom and photo-mechanical imagery

By the 1950s printmaking was moribund. In a special 'Prints' issue of *Art News* (in 1972), Thomas Hess recalled that in about 1955 he asked the painter Franz Kline whether he was interested in making prints, to which the reply was: 'No, printmaking concerns social attitudes, you know – politics and a public . . . like the Mexicans in the 1930s; printing, multiplying, educating; I can't think about it; I'm involved in the private image.' By 1970 the situation was transformed; the artist who was not involved in printmaking was the exception. One painter remarked to Hess: 'In the '50s, when you saw a friend on the Long Island Railroad on an early Wednesday morning, you knew he was going to town to see his shrink. Nowadays, you know he's on his way to work with his lithographer.'[24] This change was brought about by a general shift in artistic attitudes away from the introspective and gestural approach of Abstract Expressionism (itself a reaction against the literalness of the photograph) towards the more public stance of the Pop artists.

Paolozzi was a leading figure in this shift in attitude in England and played an important part in the revival of interest in printmaking. His interest in collage was aroused by some Dadaesque examples, which he saw on a visit to Paris in 1947–49. He began to make collages himself, using his favourite images from American pulp magazines which he had collected since he was a child. American publications of the 1940s seemed exotic in comparison with staid British ones. The glossy extremes of consumerist imagery, Science Fiction fantasy and grotesque horror-comics provided a harsh contrast to Britain in the grip of austerity. Brought over as ballast, the pulps presented a fantasy world made from photographic and photo-process images. This form of photo-mechanical printmaking, unfettered by British standards of good taste, was as far removed from the refined whimsy of a mediocre English lino-cut as could be imagined. Not content with making collages out of this riot of imagery, Paolozzi projected it as slides in an astounding lecture at the ICA in 1952. The audience was the small Independent Group of painters who formed an avant-garde within the ICA. He made no overt comment or value judgement on the images, but nor did he dismiss them as unworthy or trivial. Evidently he and other artists were coming to feel that these images reflected the world, and that they and the technology that produced them could no longer be ignored.

Richard Hamilton shares with Paolozzi an interest in all aspects of modern technology. In 1955 he and Reyner Banham organised an exhibition entitled 'Man, Machine and Motion'. It was composed mostly of photographs of machines 'made by man to extend and adapt his own organism'. This theme was prophetic of the encouragement that Hamilton has given to printmakers to extend their means of expression by the use of mechanical aids, and foreshadowed his decreasing reverence for the manual achievements of the artist as against the capacities of the machine and of the technician directed by the artist.

In 1956 he contributed to the Whitechapel Art Gallery's exhibition 'This is To-morrow' which consisted of a series of twelve environments produced through the collaboration of a painter, a sculptor and an architect. Hamilton's section explored the ambiguities of perception through the juxtaposition of popular imagery and contained all the major elements of Pop art. His collage *Just what is it that makes to-day's homes so different, so appealing*', made for reproduction as a poster and in the catalogue, is composed of carefully selected photo-mechanically printed images from magazines, including pin-ups and an early satellite picture of the earth. These form a pictorial answer to the question posed in the title. Subsequently photographic images have

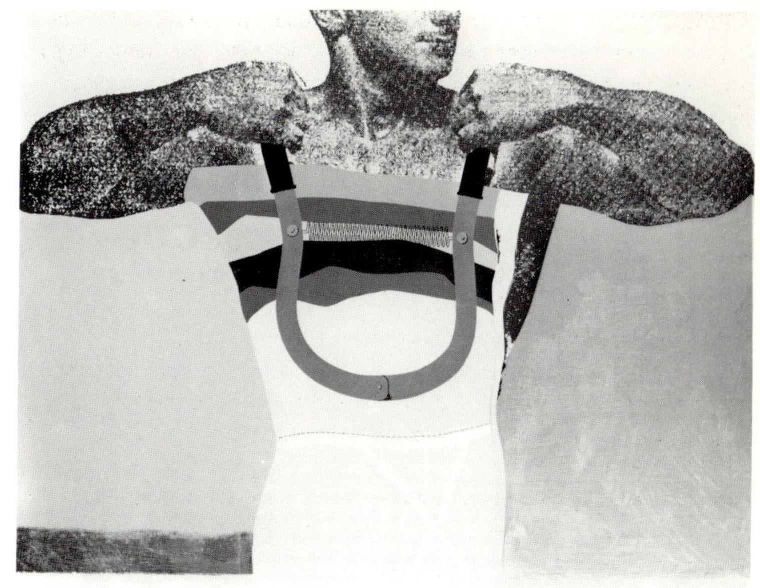

become an integral part in both Hamilton's paintings and prints (Plate C). In his own words 'the directness of photographic techniques, through half-tone silk screen for example, has made a new contribution to the media of painting'.

Independently in America, artists had been returning to figurative work, but their chosen subject matter would have seemed unspeakably banal to earlier generation of painters. In 1955 Jasper Johns made his American Flag paintings. The same year Rauschenberg produced *Bed* which consisted of paint applied to his own bedding (pillow, sheets and patchwork bedspread) attached to a stretcher. Shortly afterwards he began incorporating photo-mechanically printed images and found objects, such as Coke bottles and car tyres, in works that he described as 'Combines'. His desire to include these familiar articles in his paintings, without breaking the picture surface, led him to experiment with photo-sensitising the surface of the canvas and, more successfully, to screenprint images on to canvas (Plate D). Andy Warhol, who had been a commercial artist specialising in illustrating shoes, made his name as a painter in 1960. By 1962 he was using photographic screenprinting, in collaboration with Aetna Silk Screen Supply Company, to transfer images of mundane objects or familiar subjects on to canvas (Plate E).

Pop art was, according to Hamilton's definition, 'Popular (designed for a mass audience). Transient (short term solution). Expendable (easily forgotten). Low Cost. Mass Produced. Young (aimed at Youth). Witty. Sexy. Gimmicky. Glamorous. Big Business.'[25] This description could and did fit the printed photographic imagery of the glossy world of advertising, which was increasingly adopted in prints intended as artistic statements. The artists, in general, disclaimed a critical or social message. Warhol denied that there was any content in his work behind or beyond what is immediately obvious: 'If you want to know about Andy Warhol just look at the surface of my paintings and films and me and there I am. There's nothing behind it.'[26] They did not even

worship the machine. The popular images just *were*; they existed, they were for use. As painters for centuries had reflected the world around them even when portraying legendary subjects, so the Pop artists reflected their world and the myths of the majority. That the world might seem in the eyes of a certain type of critic to be gross, materialistic and imbued with the fatuous non-values of the advertising agency was irrelevant. Pop art accepted and reflected the world without setting up an elitist duality between good and bad taste, artistic and inartistic images (Plates F, G, H, J). It is ironic that the familiarity and wide currency of advertising images, which had formerly prejudiced artists against photography, should now attract the most progressive painters. It is interesting, furthermore, that painters made use of photographic printing techniques to transfer images to their canvases, before those techniques were widely taken up by 'original' printmakers.

Screenprinting was the ideal medium for the Pop artist. It was developed early in the twentieth century, and its simplicity quickly made it popular among signmakers and advertising firms. Its potential as a fine art medium was first explored in the USA in 1938. At this time the term 'serigraph' was invented to distinguish its use as a means for original expression from its commercial application. But leading artists did not take up the technique with enthusiasm until the 1960s, when they found it attractive because of its commercial background and its consequently advanced technology. Photographic techniques for screenprinting were first patented in 1906. Since the 1920s photo-mechanical screenprinting had been used to create the kind of images

Fig. 14   Colin Self, *Power and Beauty No. 3*, 1968. Etching

The bulbous lines of the car are rendered by massive enlargement of the half-tone dots of the source image. The artist has exploited an inherent characteristic of the etching process, which at this extreme results in embossing, giving a sculptural feel to the surface. The set was published by Editions Alecto, which aims to provide facilities for artists to produce multiples of their work in a variety of media.

▷ Fig. 16   Michael Rothenstein, *Clench*, 1969. Screen and relief print

In his book, *Frontiers of Printmaking* (1966), Rothenstein commented on the tension created when craft and industrial techniques confront one another. This print demonstrates this point and illustrates his belief that 'photography is where all media meet'. Pincers, screenprinted from a photograph, are shown withdrawing a steel foil clench from a piece of wood, relief printed from a found piece of timber. The subject relates to the artist's earlier sculptural works, using smashed crates.

Fig. 15   Jim Dine, *Smoke in a vice*. From the set 'Tool Box', 1966. Screenprint and collage on matt acetate

This set was printed and assembled at the Kelpra Studio and published by Editions Alecto. It consists of images from industrial design magazines and old fashioned engineering textbooks wittily combined with drawn marks. In this print, the basic concept of a machine having a tangible grip on something as insubstantial as smoke is emphasised by the matt, translucent nature of the polished acetate film. Later the artist wrote off this set as mere reproductions, avowing that printmaking was for him traditional . . . 'deeper, more profound and handmade'. (Quoted from Pat Gilmour, *The Mechanised Image*, Arts council, 1978, p. 110.)

to which the Pop artists turned in their search for a way back to figurative subject matter. By 1960 it had been developed into a sophisticated process, capable of translating the most diverse and colourful images into thin or thick, transparent or opaque, layers of ink. The first artist in England to use screenprinting as an original medium was Gordon House. In 1961 he commissioned a commercial painter, Chris Prater of the Kelpra Studio, with whom he had worked as the graphic designer on a number of Arts Council posters, to print his abstract grid *Peach Square*. In 1962 Paolozzi approached Prater to reproduce some of his early collages, but soon realised just how many more possibilities for new work the process offered (Plate A). In the same year Richard Hamilton persuaded the ICA to commission artists, mainly connected with the Institute, to make screenprints at Prater's studio. The fruits of this project, which resulted in 24 artists producing one print each, in editions of 40, were shown in the ICA gallery in 1964. Some of these artists never made prints again but for others it was a turning point: both R. B. Kitaj (Plate K) and Joe Tilson (Plate N) have gone on to make printmaking a major part of their artistic activity. The exhibition renewed the interest of art publishers in this field who were encouraged by the public's demand for relatively inexpensive works by fashionable artists. Editions Alecto (Figs. 14, 15) and the Marlborough Gallery were among publishers to commission prints at this time.

Photo-mechanical printmaking as a vehicle for thought processes

Photographic techniques infiltrated the original printmaking world in order to allow the incorporation of photographic imagery in prints (Fig. 16). More recently there has been a movement in art away from this kind of representation towards the exploitation of the procedure of creation. Many artists have questioned the value of traditional painting and sculpture, emphasising the thought behind their work rather than its visual impact. Film and video have been used widely as a means of making art about the artist's performance. Photography has become part of the exploration of what art is. Its use in printmaking has followed these general trends, this particular combination making it possible to convey concepts which cannot be expressed by either photography or traditional printmaking alone.

In some cases printmaking has provided a cheap means of multiplying a photographic record of an event. It is the event that constitutes the work of art, and so the print, as a record, bears a similar relationship to the event as a reproductive print after a painting bears to its original (Figs. 17 & 18; Plate P). In other examples photographs, or photographic sequences (Fig. 19; Plates L, T, R) are central to the print but the work has been executed in etching, screenprinting or lithography because these techniques allow the artist to give the image visual qualities unobtainable by other means. In yet other prints, especially where found objects form the image, printmaking has provided the most suitable means of creating a multiple which expresses the desired associations (Plate Q). In Broodthaers' *La Soupe de Daguerre* (Plate S), printmaking is itself one of the associations. Under the stuck-on photographs of vegetables is a screen printed mock-label, which implies an equivalence between the hotch-potch of printing techniques and the multiple ingredients of vegetable soup.

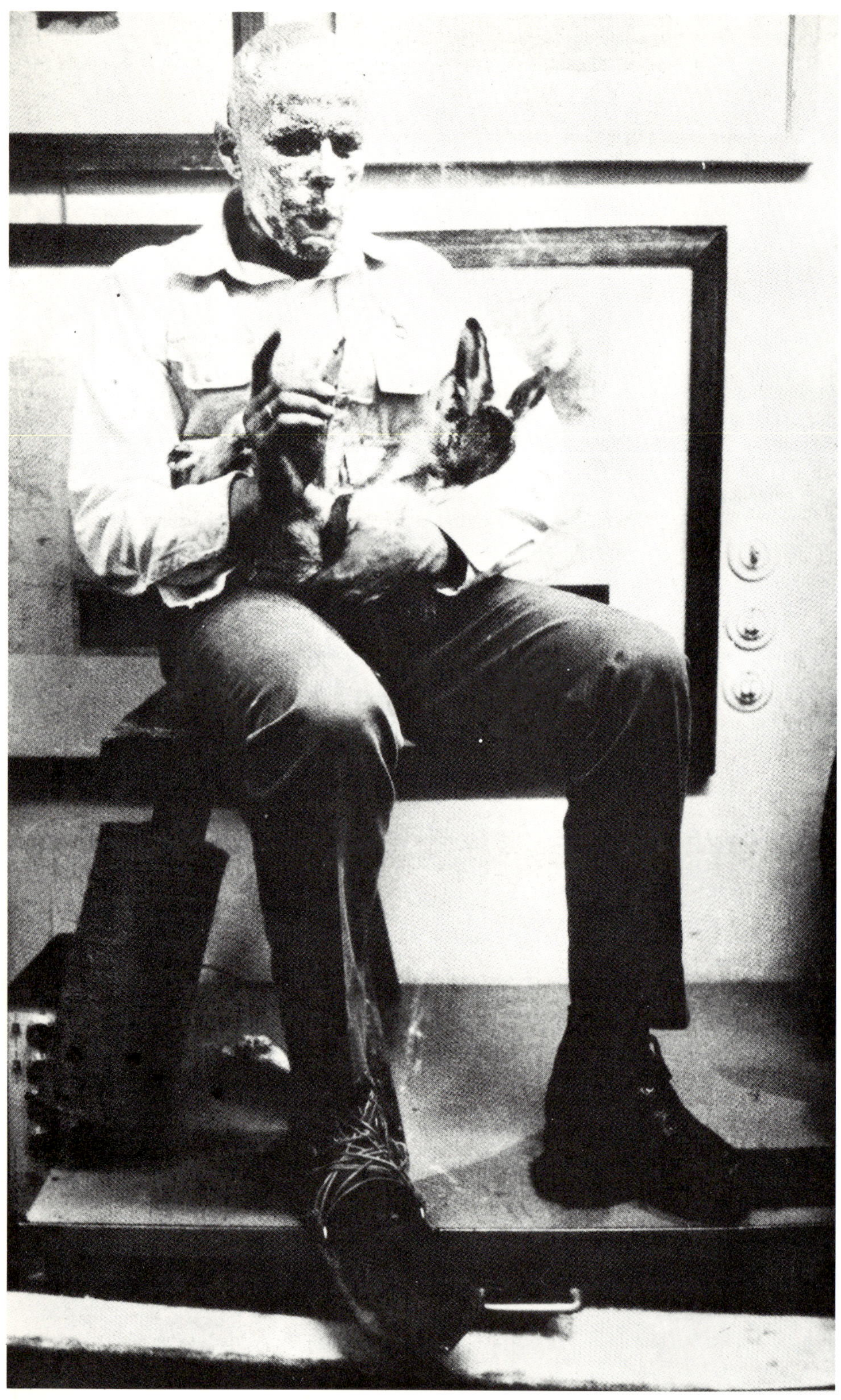

A record of an action in the Galerie Schmela,
Düsseldorf, reproduced cheaply by a commercial printing process.

Fig. 18   Michael Craig-Martin, *Untitled*,
1970. Offset lithograph

This print shows the development of a
minimalist structure and is a permanent
record of it.

▷ Fig. 19   Gail McKennis, *Janet and the
Vermeer II*. From the series 'Art of the Real',
1972. Lithograph

This print plays with ideas about reality. A
reproduction of a Vermeer is juxtaposed with
a photograph of a girl. Vermeer used the
camera obscura; the photographic reproduction of the painting increases and restores its
original 'photographic' quality. Obversely the
elimination of tones in the photograph of the
girl, depicted looking out through a spattered
border, gives this image the quality of a time-
darkened painting.

Photography as a tool

An important part in the recent production of prints has been played by technicians. In America Tatyana Grosman of Universal Limited Art Editions bought her first lithographic press in 1957 and began trying to interest major artists in printmaking. Most artists were sceptical. Rauschenberg has said: 'I began lithography reluctantly, thinking that the second half of the 20th century was no time to start writing on stones.'[27] Johns's initial reaction was no more enthusiastic, although he was quickly converted. He did not regard printmaking merely as a means of multiplying painted images: 'Paintings and prints are two different situations. . . . Primarily it's the printmaking techniques that interest me. My impulse to make prints has nothing to do with my thinking it's a good way to express myself. What interests me is the technical innovation possible in printmaking.'[28] To explore technical innovations, many of which depended on skill in photographic processing, artists had to seek the help of technicians. A precedent for this kind of participation in an artistic statement has been set by Moholy Nagy in the twenties when he ordered an enamel painting from a commercial firm over the telephone simply by giving the colour references and dimensions. In 1965 Kenneth Tyler set up Gemini GEL in Los Angeles with the aim of providing technical support, in the form of expertise and equipment, for artists in a wide range of media. In England, Stanley Jones of the Curwen Studio provides this kind of service in lithography and has developed the commercial Diazo process as a means for original expression. Cliff White's White Ink Ltd, specialising in the intaglio processes but also doing some screenprinting, was set up in 1970 while Chris Betambeau split from Kelpra to form Advance Graphics in 1968. Chris Prater often uses the metaphor of the composer-conductor-orchestra to describe the relationship between the artist, chief technician and the workshop printers, seeing himself 'as the conductor, rendering faithfully the composer's imagery'. 'A lot of the success is in understanding the artist . . . when he says green, you must know exactly what shade of green is meant . . . it means knowing the artist's work as well as his way of working.'[29]

Victor Pasmore writing on the relationship between the artist and printer sees the intervention of the camera as the distinguishing feature of contemporary methods as compared with traditional ones: 'The big difference is in the use of photography in connection with a preliminary maquette as an intermediate process between artist and printer. This obviates the necessity of the artist making the whole print directly on the plate while at the same time it enables the printer to reproduce from the maquette the actual 'handwriting' of the artist. In this new context the work of the printer is to clarify those parts not picked up properly by the camera or to make any subsequent modifications required by the artist. Likewise the role of the artist is to re-draw or add directly on the plate any personal motifs which cannot be copied by the printer. The new relationship is very similar to that carried out by the Japanese in the famous woodcuts of the Ukiyoye school; but with the difference that the camera has been brought in as a connecting link. In the Japanese woodcut the woodcutter and printer had to copy the work of the original artist, as well as interpret it in woodcutting form. In the present form the printer copies and interprets the work only partly in those places where either the camera failed to, or where the artist's personal handwriting is not needed.'[30]

Once the use of photographic techniques became accepted, they were quickly taken up as a tool and used in the making of prints which do not necessarily appear to have a photographic base. Photography has made it possible to build up very complex images quickly, in a way that is not possible by hand. It enables the artist to juggle with the scale and position of the images, before using a traditional printmaking technique to reproduce and unify the final image.

Sometimes photography is simply used to transfer a hand-drawn image on paper to a plate, screen or stone, which may then be worked on by the artist to a greater or lesser degree (Plates, V, U). Henry Moore draws on transparent film through which the image is transferred photographically to a photosensit-ised aluminium lithographic plate. The surface of the transparent film offers no resistance to the drawing implement and the photographic process allows the artist to draw the image the same way round as he intends it to be printed; the technique allows the full autographic quality of his drawing to be expressed in ·the completed lithograph. If the artwork is made on paper, additional work can be added or subtracted and corrections can be easily made on the photographic film before it is used to transfer the design to the printing surface. Or, as in the case of many of Paolozzi's screenprints, it can be the support for substantial amounts of original work. In addition photography has the advantage that it allows images to be 'recycled' (Paolozzi's word). He has torn-up his collages and re-arranged the forms in order to create new images. Henry Moore re-combines earlier images as the background to new lithographs by cutting and re-using the film on which the drawing has originally been made.

Photography has given the printmaker a new range of found and accidental effects (Fig. 20). Some prints take a photographic image as their starting point but the image is so enlarged, reduced, obliterated or otherwise manipulated that it is the patterns or textures created by the photographic techniques rather than the subject of the original photograph that are central to the print. Harold Cohen's series of portraits of Richard Hamilton (Plate M) exploiting the effect of the magnified dot, which results in an increasingly abstract image, are well known. Hamilton himself has created an abstract pattern through the enormous magnification of a holiday snap.

In addition the selected image, whether of photographic origin or not, can be subtly or grossly manipulated by photographic techniques (Plate Z). With their aid, Gerd Winner (Plate W) transforms straightforward photographs of buildings into striking screenprints. Using sophisticated posterisation techni-ques, which allow infinite variations in the build-up of colour, tone, surface and registration, Ivor Abrahams changed the often poor quality illustrations from books and newspapers, which were his source of material for the series inspired by Poe's *Tales and Poems*, into commanding and disturbing images. Even the simple translation of a photographic print such as Alyson Hunter's *Window* (Plates X, Y) into the soft rich tones of photogravure can produce unexpectedly satisfying results.

The use of photo-mechanical techniques in printmaking has coincided with a creative approach to the medium rarely experienced since the invention of photography. It has allowed the incorporation of a new kind of subject matter and increased the range of visual effects available. The result has been the

In this print, Tillyer used photography to give a hand-controlled etching mechanical sharpness. A letratone grid was photographically enlarged and screened on to the metal plate but the width of the line, dependent on the length of its immersion in acid, was controlled intuitively. The letratone grid gives the image a precision unobtainable with hand-ruled lines.

production of vivid, often garish, reflections of the world in techniques in tune with the imagery, which is particularly apt for mass-produced artistic statements. Moreover, the removal of much of the drudgery and technical expertise to the realm of the printer has given the artist more freedom to grapple with the content of his art. In 1914 Marcel Duchamp selected his first readymade, a rack for drying bottles purchased from a Paris department store. In 1917 he exhibited a urinal. Placing it upside down and signing it, he gave it the title *Fountain*. His much quoted justification for this exhibit was that: 'Whether Mr Mutt with his own hands made the fountain or not has no importance. He CHOSE it. He took an ordinary article of life, placed it so that its useful significance disappeared under the new title and point of view – created a new thought for that object.'[31] And in connection with the *Bottlerack* he said his work aimed to 'reduce the aesthetic consideration to the choice of the mind, not to the ability or cleverness of the hand . . .'.[32] The eventual whole-hearted acceptance of this attitude in printmaking has opened up new vistas for the art form.

Notes

1. This phrase recurs throughout *Prints and Visual Communication*, published by Routledge and Kegan Paul, London, 1953.
2. Philosophically, photography could be regarded as a branch of printmaking, a kind of latent ink process printing, using silver salts rather than carbon based printing inks.
3. For an example of O. G. Rejlander's work see the periodical *The Illustrated Photographer*, 1869.
4. Letter in the *Guardian*, 6 March 1967.
5. *Printmakers Council Handbook of Printmaking Supplies*, edited by Silvie Turner, 1977.
6. *The Magazine of Science and School of Arts*, 27 April 1839. There are similar photogenic drawings on woodblocks in *The Mirror of Literature, Amusement and Instruction*, 20 April 1839.
7. J. Adhémar, *Twentieth Century Graphics*, 1971, p. 10.
8. Letter to Leonard Smithers, 15 January 1898. Quoted from *The Letters of Aubrey Beardsley*, edited by Henry Maas, 1971.
9. Ibid. 26 December 1897.
10. Ibid., 10 December 1897.
11. This method had been pioneered by English artists, such as Havill and Wilmore, in the 1830s.
12. This is particularly true in America. A contemporary cliché verre school, including work in colour, has grown up at Wayne State University, Detroit.
13. W. P. Frith, *My Autobiography and Reminiscences Vol. III*, Richard Bentley and Son, London, 1888.
14. J. Adhémar, op. cit. p. 10.
15. In fact Francesco Bartolozzi and Richard Yeo were elected but the former was elected as a painter and the latter as a sculptor.
16. Sidney C. Hutchison, *The History of the Royal Academy 1768–1968*, 1968, pp. 89–90.
17. F. P. Marinetti, *Selected Writings*, edited by R. W. Flint, 1972.
18. Hans Richter, *Dada, Art and Anti-Art*, 1965, p. 117.
19. Ibid.
20. Ibid. p. 116
21. Ibid. p. 159
22. Jan Tschichold himself put the case for photography and new typography in his article 'New life in print', *Commercial Art*, 1931 Vol. IX, p. 2.
23. Hans Richter, op. cit. p. 118.
24. T. Hess, 'Prints: where history, style and money meet', *Art News*, 1972, p. 29.
25. A letter of 16 July 1957 to Alison and Peter Smithson in reply to a discussion about 'This is To-morrow'.
26. Simon Wilson, *Pop*, 1974, p. 18.
27. The Minneapolis Institute of Arts, Exhibition Catalogue, *Robert Rauschenberg: Prints 1948/1970*, 1970.
28. Diane Kelder, 'Tradition and craftsmanship in modern prints', *Art News*, January 1972, p. 59.
29. Arts Council of Great Britain, Exhibition Catalogue, *Kelpra Prints*, 1970.
30. Bryan Smith, 'Victor Pasmore and the Printer', *Penrose Annual*, 1974.
31. Art Institute of Chicago Exhibition Catalogue, *Idea and Image in Recent Art*, 1974, p. 10.
32. Ibid.

Bibliography

General

Ades, Dawn; *Photomontage*, 1976
Castleman, Riva; *Prints of the Twentieth Century:
a History*. 1976
Coke, van Deren; *The Painter and the Photograph*,
Albuquerque, New Mexico, 1964.
Compton, Michael; *Art since 1945*, Milton Keynes, 1976.
Eder, Josef M.; *History of Photography*, New York, 1945.
Eichenburg, Fritz; *Art of the Print*, 1976.
Finch, Christopher; *Pop Art: Object and Image*, 1971.
Gernsheim, Helmut and Alison; *The History of Photography*,
1969.
Gilmour, Pat; *Modern Prints*, 1971.
Godfrey, Richard T.; *Printmaking in Britain*, 1971
Ivins, William; *Prints and Visual Communication*,
Cambridge, Mass., 1953.
Jussim, Estelle; *Visual Communication and the Graphic Arts, 1974.*
Kostelanetz, Richard (ed); *Moholy-Nagy*, 1971.
Lippard, Lucy R., and others; *Pop Art*, 1966.
Newhall, Beaumont; *The History of Photography*, New
York, rev. ed. 1964.
Richter, Hans, *Dada, Art and Anti-Art*, 1965.
Russell, John, and Gablik, Suzi (eds); *Pop Art Redefined*,
1969.
Scharf, Aaron; *Art and Photography*, 1968.
Selz, Peter; *John Heartfield: Photomontages of the Nazi
period*, 1977.
Smith, Edward Lucie; *Movements in Art since 1945*,
1972.
Sontag, Susan; *On Photography*, 1978.
Spencer, Charles (ed.); *A Decade of Printmaking*, 1973.
Thomas, D. D. B.; *The First Negatives*, 1965.
Wakeman, Geoffrey; *Victorian Book Illustration*,
Newton Abbot, 1973.

Walker, John A.; *Art since Pop*, 1975.
Wilson, Simon; *Pop*, 1974.

Technical

Biegeleisen, J. I. *Screenprinting: a contemporary guide
to the technique of screenprinting for
artists, designers and craftsmen*, 1974.
Brunner, Felix; *A Handbook of Graphic Reproduction
Processes URICH, Switzerland.*
Vicary, Richard; *The Thames & Hudson Manual of
Advanced Lithography*, 1978.

Exhibition Catalogues

Art Institute of Chicago, *Idea and Image in Recent
Art*, 1974.
Arts Council of Great Britain, *Dada and Surrealism
Reviewed*, 1978.
Kelpra Prints, 1970.
*The Mechanised Image: an historical perspective
on twentieth century prints*, 1978.
*Eduardo Paolozzi: Sculpture, drawings, collages
and graphics*, 1976.
Photo-Realism, 1974.
Pop Art, 1969.
Davison Art Center, Wesleyan University, Middletown,
The Prints of Richard Hamilton, 1973
Minneapolis Institute of Arts, *Robert Rauschenberg:
Prints 1948/1970*, 1970.
Tate Gallery, London, *Richard Hamilton*, 1970.
Victoria & Albert Museum, London, *"From Today Photography
is Dead". The Beginnings of Photography*, 1972.
*The Complete prints of Eduardo Paolozzi: Prints,
Drawings, Collages, 1944–77*, 1977

Plates

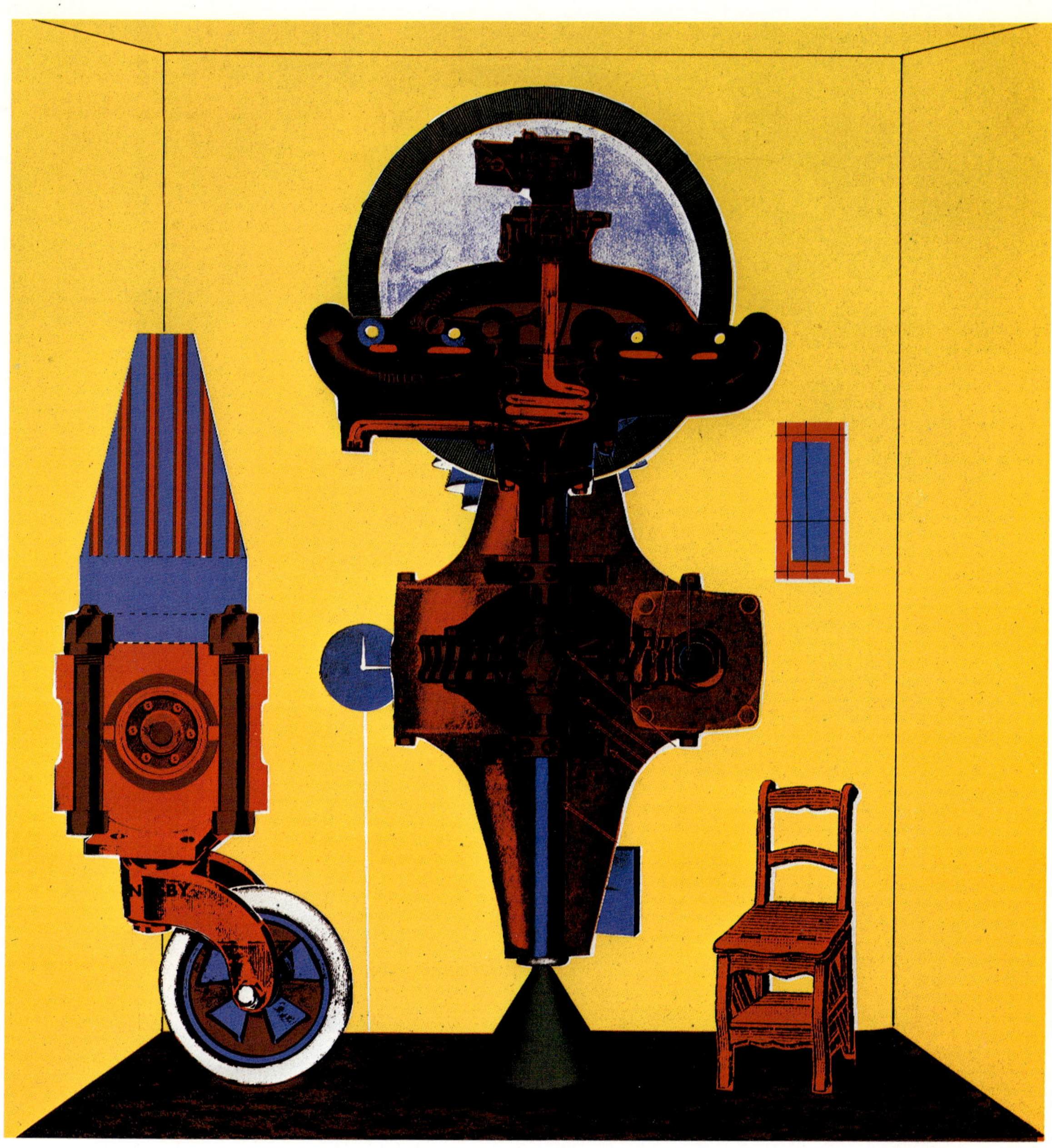

Plate A   Eduardo Paolozzi, *Metalisation of a Dream*, 1963.
Screenprint

Paolozzi had had several of his collages reproduced by screenprinting before this date, but this was the first to be specifically designed for the medium. The Ernst-like quality of the image is appropriate; just as Ernst pioneered the medium of photomontage, so Paolozzi was among the first to use photo-mechanical printmaking creatively. The title refers to Marinetti's Futurist phrase 'The dreamt of metalisation of the human body'. Printed with kind permission of the artist.

STAR KING ROBO
スター・キング・ロボ

Plate B   Eduardo Paolozzi, *Hollywood Wax Museum*. From the set
'Zero Energy Experimental Pile'. 1969–70. Screenprint

The image is built up from a conglomeration of the ephemera of a
consumer society. Photography is shown as the creator and
disseminator of garish and violent images which reflect the more
unpleasant aspects of daily life and the ugly side of technology. The
technical bravura of the method of the print's production, with the
images screenprinted on to thin astralux paper sealed in laminated
plastic, is appropriate to its mechanised imagery. Printed with kind
permission of the artist.

Plate C   Richard Hamilton, *My Marilyn*, 1965. Screenprint

Hamilton has always been fascinated with the special qualities of a
photograph and how they relate to painting and printmaking. Of this
print he said 'Since the basis of the print was a group of photographs
with markings made by Marilyn Monroe's hand, it became an
objective of the prints to produce a painterly result without actually
making marks with mine'. Photographic images, some of them in
negative, are presented in painterly colours giving an effect
unobtainable by other means. (Quoted from Davison Art Center,
Middletown, Connecticut, Exhibition Catalogue, *The Prints of
Richard Hamilton*, prepared by Richard S. Field, 1973, No. 15.)

◁ Plate D   Robert Rauschenberg, *License*, 1962. Lithograph

Shortly after making his first group of lithographs, of which this is one, Rauschenberg recalled the reluctance with which he had taken up the medium, going on to say: 'This biased idea was soon consumed in the concentration any unfamiliar medium requires. Lack of preconception and recognition of the unique possibilities in working on stone, not paper or canvas, suggested that the approach acknowledge this. As part of its content, my work has always had to include and utilize actual elements from everyday life which were not necessarily considered artist's materials. . . My lithography is the realisation and execution of the fact that anything that creates an image on stone is potential material. The image that is made by a printer's mat, a metal plate, a wet glass or a leaf plastically incorporated into a composition and applied to the stone, stops functioning literally with its previous limitations. They are an artistic recording of an action as realistic and poetic as a brush stroke.' (Quoted from the Minneapolis Institute of Arts, Exhibition Catalogue, *Robert Rauschenberg: Prints 1948/1970*. 1970.)

△ Plate E   Andy Warhol, *Marilyn Monroe*. From 'Ten Marilyns,' 1967. Screenprint © by V.A.G.A, New York, 1979

One of a series of ambiguous and disturbing images, printed in ten different colourways in a range of crudely modulated garish colours. It is left to the spectator to speculate as to why Warhol chose so to distort the original photograph. The cosmetic-like areas of colour printed out of register on top of the enlarged image are reminiscent of poor quality magazine pictures.

Plate F   Peter Phillips, *Lion*. From the series 'Pneumatics', 1969

The sixties obsession with slickness and styling is reflected in the subject and finish of this high quality, multi-coloured photo-mechanical print.

Plate G   Peter Blake, *Beach Boys*, 1964. Screenprint

This vivid interpretation of the pop group standing by their limousine was based on a press photograph and was Blake's contribution to the ICA project.

THE
BEACH
BOYS

Plate H   Allen Jones, No. 2 from the series 'Life Class', 1968.
Lithograph and photo-lithograph

Jones is fascinated by the tension of male and female sexuality as
expressed in fetish magazines and here explores the subject by
juxtaposing commercial process work with a painterly image. The top
half of the print was drawn by hand, while the lower half reproduces a
photograph of the legs of a model.

Plate J   Richard Smith, *Zoom*, 1964. Screenprint

This print, commissioned by the ICA, can be interpreted as a
commentary on advertising. The strange prespective of the cigarette
packet, achieved with the aid of photography and alluded to in the
title, gives the image a menacing quality.

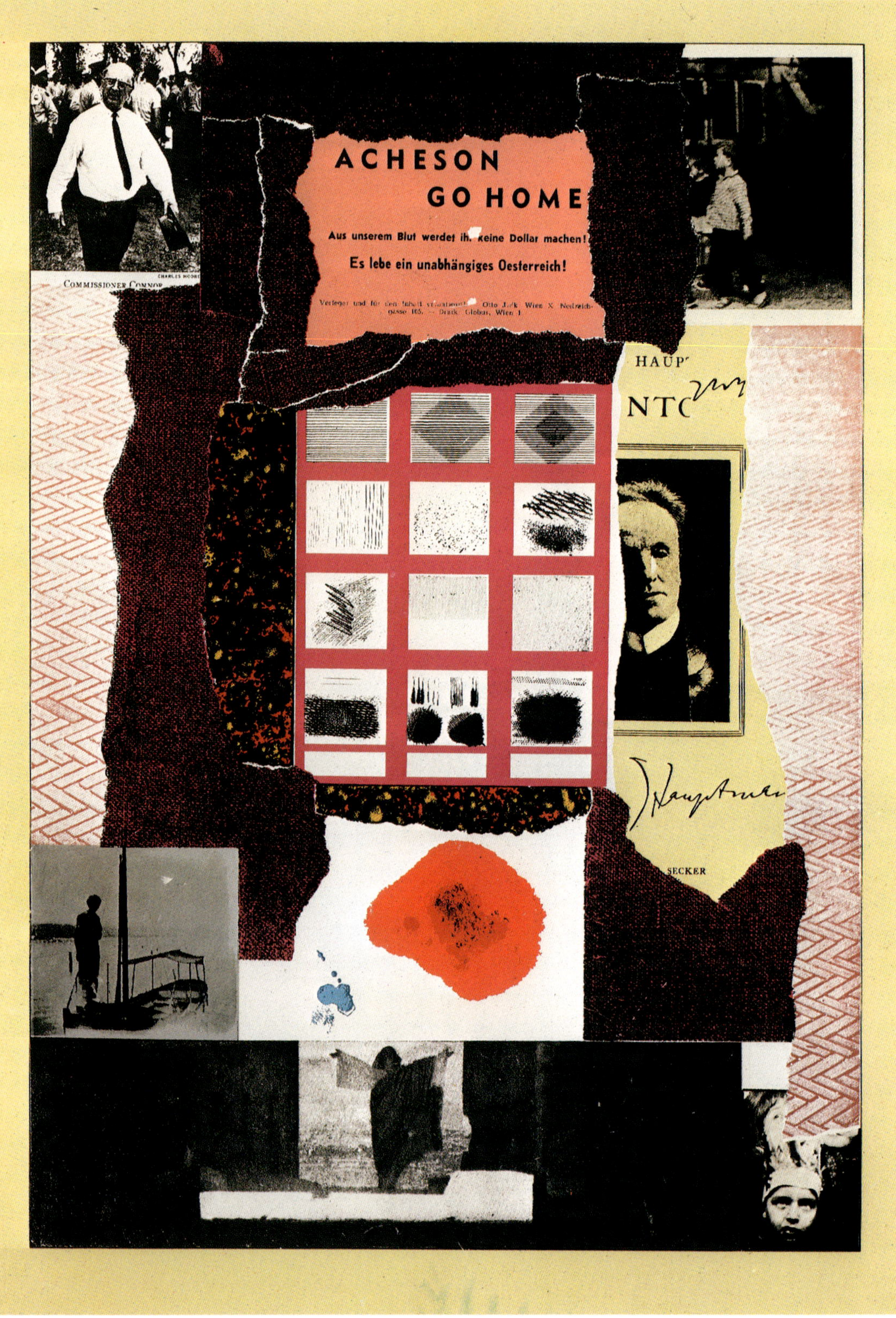

COMMISSIONER CONNOR
ACHESON
GO HOME
Aus unserem Blut werdet ihr keine Dollar machen!
Es lebe ein unabhängiges Oesterreich!
HAUP
NTC
SECKER

Plate K   R. B. Kitaj, *Acheson Go Home*, 1963. Screenprint

Although Kitaj is said to have been reluctant to take up printmaking, he quickly realised the potential of screenprinting. In this, his first print, he uses the printing process to integrate a diverse combination of images including a newspaper cutting about the visit of Dean Acheson to Austria and a spot of real blood splashed on to the original collage. The allusiveness and intellectuality of the images and their juxtaposition, the appeal of the work as an abstract design, and the sophisticated use of printing technology are characteristic of his later prints.

Plate L   Nils Burwitz, *Namibia: Heads or Tails*, 1979. Screenprint

This political statement by a South African artist is a technical tour de force. The texture of the original iron notice, with the enamel peeling away where bullets have damaged the surface, is reproduced in three dimensions by the use of especially thick inks. Unexpectedly, the back of the sign is represented on the back of the print.

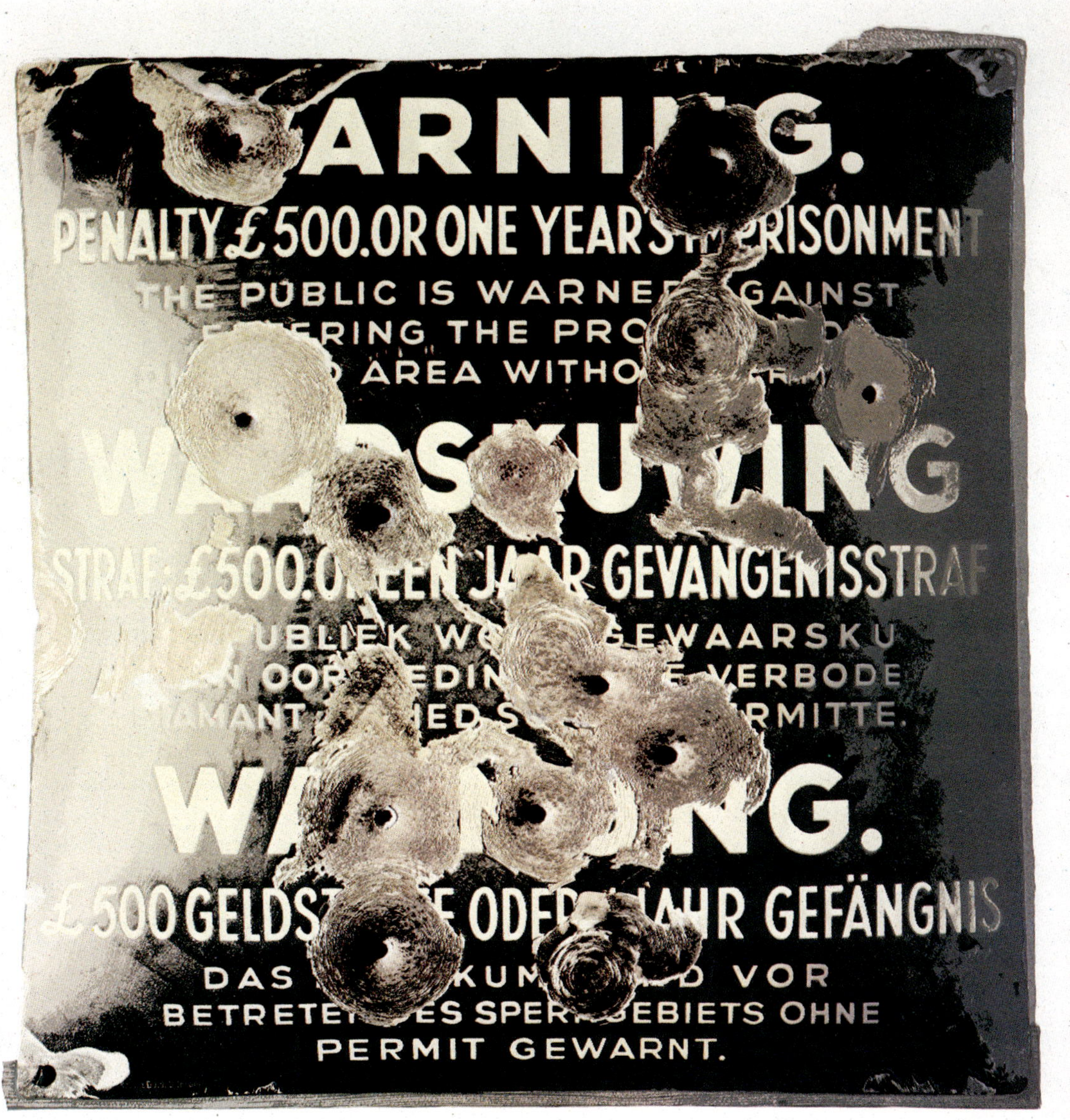

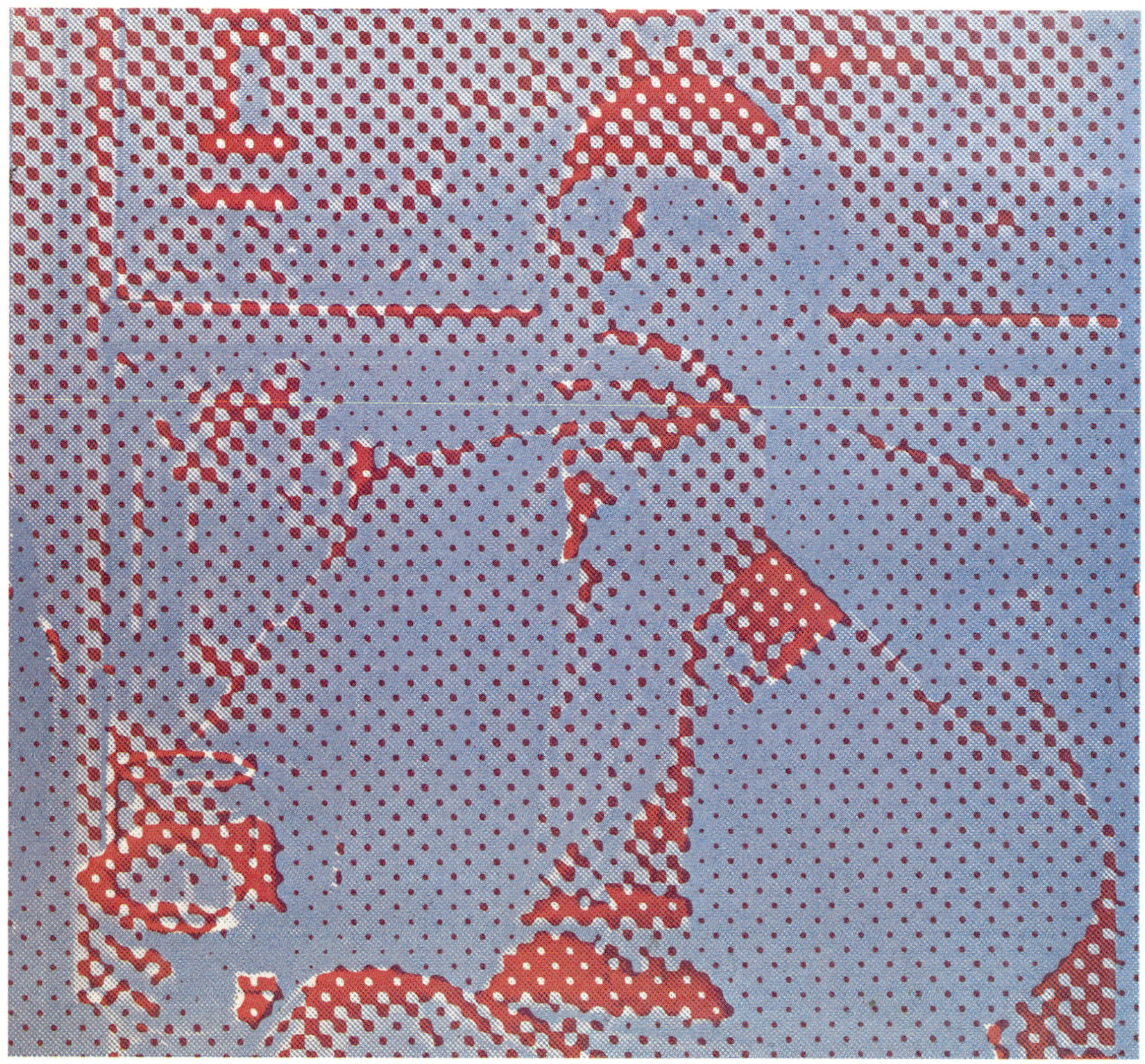

Plate M   Harold Cohen, *Richard III*, 1967. From a set of
screenprints

This print has the half-tone dot structure of photo-process printing as
its subject. A screened photograph of the artist, Richard Hamilton, is
enlarged progressively, creating a set of prints which become
increasingly abstract as the dot structure grows and separates.

Plate N   Joe Tilson, *Transparency Clip-o-matic Lips*, 1967.
Screenprint on plastic film

In his contribution to the Arts Council catalogue to the exhibition of
Kelpra prints held in 1970, Tilson recalled: 'Talking to Chris I
discovered that the way he worked was divided into photographic
and hand-cut stencil processes – both essentially commercial in their
development as opposed to what I had seen in art schools with
crayon tusche on the screen. So I started work on an idea using
photographs of pilots and aeroplanes repeating and incorporating a
coarse, dotted screen to give it the look of a printed page, and
combined this area of the print with painted and cut paper and 'zip-a-
line' tape as a basis for translation into hand-cut stencils. This
process combining both areas I still use.' This large print glories in
the frank photographic methods of production and the luscious
nature of the original photographic transparency by retaining the
slide format of the original image.

Joe Tilson 1967.
56/70.

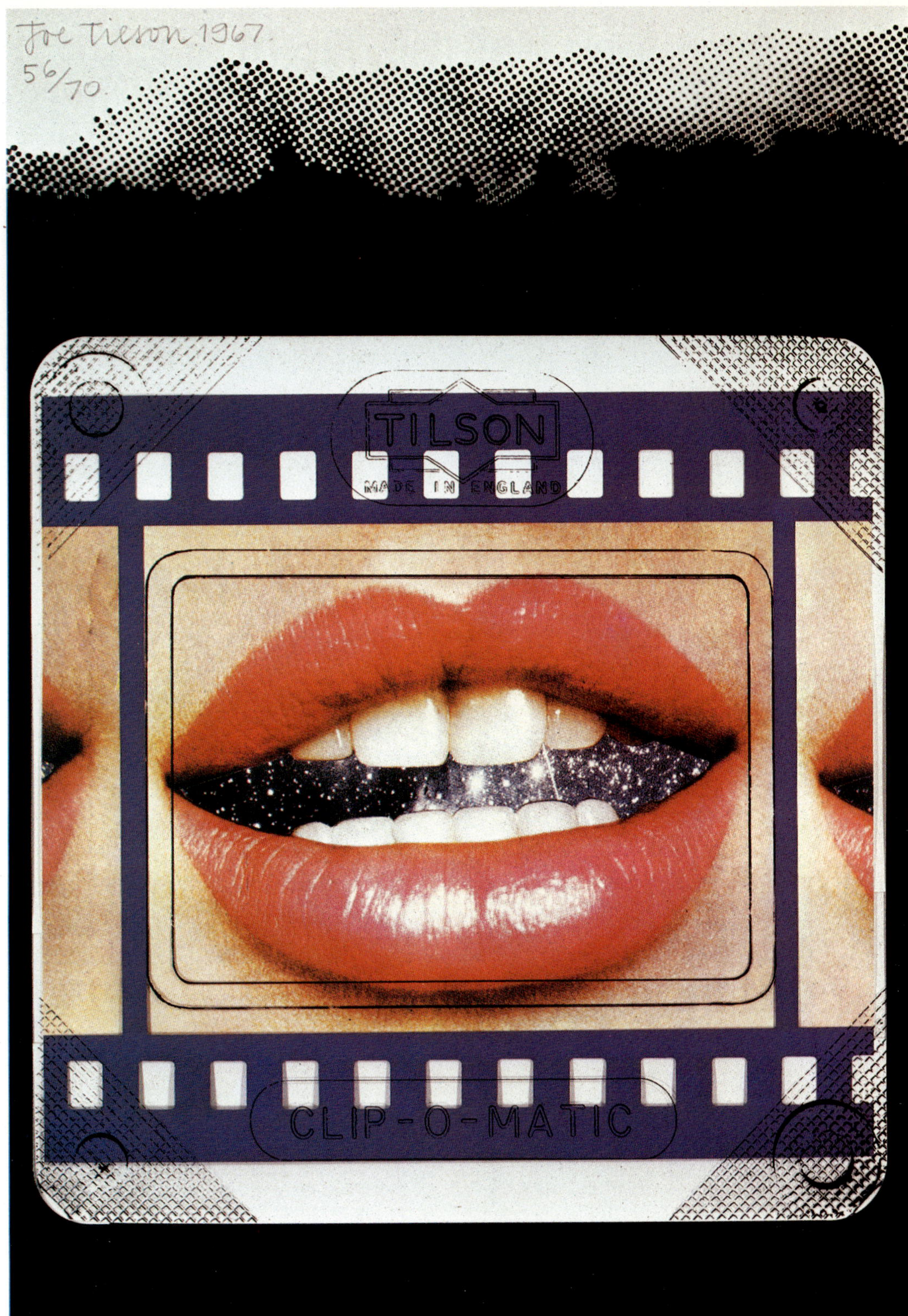
TILSON
MADE IN ENGLAND
CLIP-O-MATIC

◁ Plate P   Christo, *Wrapped monument to Leonardo*, 1970. Offset lithographs

Two prints which record the immediacy of an art event. The first is a record of the artist's plans using drawing and a photograph, and the second records the result.

▷ Plate Q   Richard Hamilton, *The Critic Laughs*, 1968. Offset lithograph, laminated, and retouched with enamel paint, and screenprint

Similar problems to those that beset photographers in the 19th century intrigued Hamilton as he observed that modern colour photographic prints fade rapidly in sunlight. In his words: 'Changes of colour occur almost immediately and a few months can show almost complete decay. The photographic stylisation of [this] print is a parody of Braun domestic appliance presentation. Such soft focus effects would be difficult to achieve with craft printing techniques. It was therefore reproduced by offset litho and laminated with clear plastic film to heighten the photographic quality. The margins were screenprinted with a matt white which makes it appear like a photograph mounted on a board . . . The interest of the print is the relationship of paint and photographic emulsion. Some thick paint was applied to the photographic print prior to reproduction and paint was added to each print by hand after lamination as well as a little collage. . . .' (ibid. No. 22.)

◁ Plate R   Tom Phillips, *Sixty-four Stop Cock Box Lids*. From the set 'The Walk to the Studio', 1978. Screenprint

This print was made from photographs of the stop cock box lids encountered by Phillips on the walk from his home to his studio. The subtle difference in the way these mass-produced and seldom noticed pieces of street furniture have weathered is emphasised by the printing process.

△ Plate S   Marcel Broodthaers, *La Soupe de Daguerre*, 1975. Colour photographs and screenprint

This is an elaborate joke at the expense of the profusion of confusing printing techniques. The luscious ingredients of vegetable soup are represented by stuck-on photographs while the mock-label is a fine piece of tromp l'oeil screenprinting.

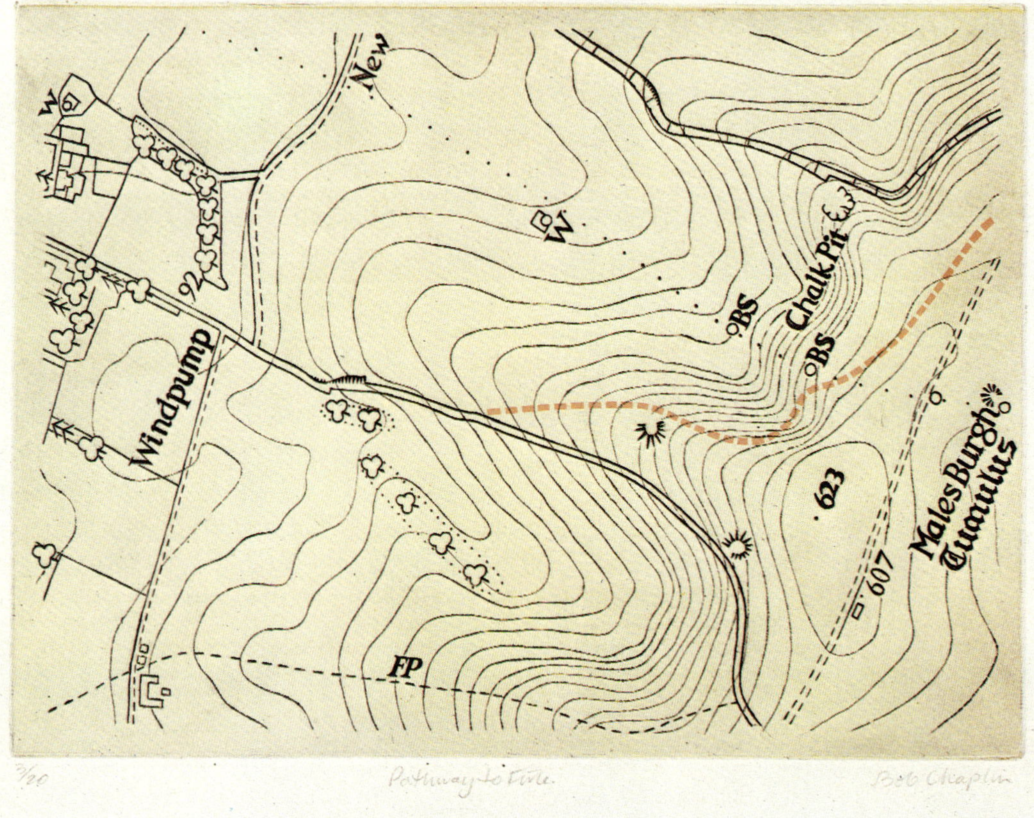

3/20   Pathway to fire.   Bob Chaplin

Plate T   Bob Chaplin, *Pathway to Firle*, 1976. Etching and aquatint

The upper photographic image, subtly softened by the aquatint ground, contrasts sharply with the diagrammatic but old-fashioned map of the same scene in the lower image. This juxtaposition gives both the romantic atmosphere of the landscape and a scientific rationalisation of it.

Plate U   Victor Pasmore, *When the lute is broken . . .*, 1974. Etching and aquatint

In the words of Cliff White at White Ink studios where this print was made: 'It is virtually impossible to re-draw the sensitive curves of an artist's line. Photography is the only way to transfer it to the plate.' The printing process gives the deceptively simple image additional depth and interest.

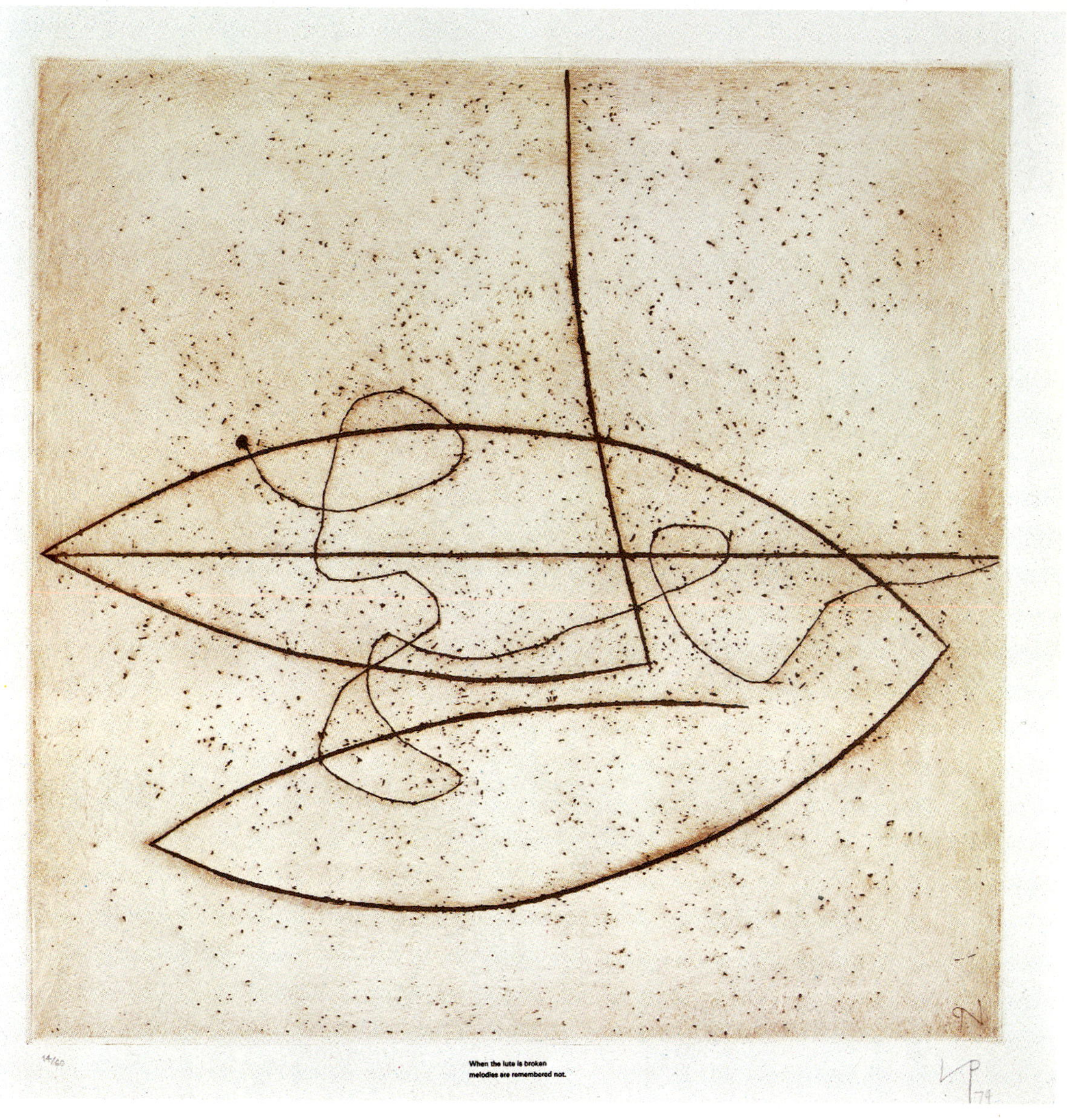

Plate V   John Walker, *Juggernaut*, 1974. Screenprint

This print has a transparency of Walker's painting, Juggernaut III, as its starting point but the photographic base has been almost obliterated by marks put on by hand. Walker drew the latter on film enabling both the photographic and the handwork to be transferred to the screen photographically, thereby giving the completed image greater homogenity. Printed in two colourways, the print shows how different effects can be achieved by printing the same screens in different coloured inks.

Plate W   Gerd Winner, *New York Canyon*, 1974. Screenprint

A photograph of drab tenements in New York is transformed into a striking image by manipulation of the printing inks. The use of posterisation, pioneered by Kelpra where this screenprint was made, allows subtle gradations in tonal density without the intervention of the commercial half-tone screen.

Plate X   Alyson Hunter, *Window '71'*, 1971–74. Photogravure

The hard edge of reality is softened by the use of commercial
photogravure, adding to the dream-like quality created by the
superimposing of one photographic image over another.

Plate Y   Donald Wilkinson, *The valley behind me*. From 'Mr Gray's Journal', 1974. Etching and aquatint

This series illustrates extracts from Thomas Gray's Journal on his visit to the Lake District in 1769. The type of the extracts is photographically reproduced from the fourth edition of West's *Guide to the Lakes* (1789) in which Gray's Journal is printed as an addenda. To make the prints, the artist visited the sites to which Gray refers, at the same time of year, taking photographs and making drawings. The photographs formed the basis of the compositions and the drawings provided colour records.

DAYTONA BEACH
RESORT AREA
10
M.P.H.

Glossary

Albumen

White of egg used in early photography for the suspension of light-sensitive silver salts. The substance was painted on to ordinary paper to form photographic paper.

Aquatint

A method of etching in tone. A porous ground is applied to a metal plate and fused to it by heat. The acid bites the plate where it is unprotected between the particles of the ground, thereby etching pits in the metal to carry the ink. The tone of any part of the printed image is dependent on the depth of these pits and so the design is built up in stages, each area being stopped out once it is adequately bitten. The process was developed in order to render more sensitively the tonal qualities of paintings in the late 18th century and was widely used for the reproduction of wash drawings and water-colours.

Autographic

That which is done by the artist's own hand. In printmaking, it often implies that the artist has prepared the printing surface himself.

Bichromate

A salt which, added to albumen or gelatine, makes these substances become insoluble when they are exposed to light. Bichromated gelatine was the essential substance for photo-mechanical work in the 19th century.

Bite

The chemical dissolving of the metal of a printing plate. A plate in an acid bath is described as being 'bitten'.

Bitumen of Judea

Asphaltum thinned with a solvent to make a varnish which hardens on exposure to light. Used by Niépce to make the first permanent photographic images. It was used in photo-mechanical processes until the end of the 19th century.

Calotype

Fox Talbot's name for his improved photographic process, published in 1841, in which a latent image is developed on light sensitised paper.

Camera Obscura

A drawing aid invented in medieval times. Originally it was a dark room in which an image of the world outside was allowed to enter through a tiny hole in one wall and fall upon a surface to be there copied or traced. A portable version was developed consisting of a box with a lens, mirror and a ground glass screen from which an image could be drawn. Eventually it evolved into the photographic camera which in all its essentials is the medieval device holding in addition the light sensitive material which makes the image permanent.

Carbon print

Perfected in 1864 by Sir Joseph Swan, it consisted of a tissue of carbon impregnated gelatine treated with potassium bichromate to make it light sensitive. Exposed under a negative the tissue hardened in proportion to the intensity of light striking it. The tissue was transferred to a temporary support and the unhardened gelatine, where less light had passed through the negative, was washed away. The image was formed from the varying thickness of the dark gelatine: the thinner the tissue at any point the lighter in tone. Superficially, carbon prints look like conventional photographic prints but they are not susceptible to fading because the carbon pigment is stable. The Autotype Co. took up Swan's process from 1866, using it to reproduce drawings and illustrations until the beginning of the 20th century. Carbon tissue was adapted to form a light sensitive resist on the surface of plates to be etched, often in combination with a line aquatint ground. It is an essential ingredient in photogravure.

Cliché verre (also known as hyalography, cliché glace, glass etching)

A photographic print made from a hand-drawn negative. The method used by Corot and Millet was to smoke a piece of glass and draw on the blackened surface so that light would travel through where the soot was removed. A piece of photographically-sensitised paper was placed underneath the glass and exposed to light. Black lines would appear on the paper corresponding to where the design had been made. Although little used in the 19th century, it has been revived recently in the USA where ingenious variants on this basic principle are being developed.

Cartes de Visite

Introduced in the late 1850s these small photographic portraits, albumen prints on card, became immensely fashionable and were often hand coloured.

Collage

From the French 'coller' to glue. The term refers to the use of any material, usually paper, glued to the surface of a picture.

Collograph

Henry Moore adapted the collotype process from its commercial use in colour reproduction to a means for original expression. He drew on a separate transparent sheet for each colour and these were then used to make collotype plates.

◁ Plate Z   Malcolm Morley, *Beach Scene*, 1969. Screenprint

Hyper-realist paintings and prints are based on photographs but normally the quality of the original photograph is elaborately imitated by hand. This print is an exception. A holiday snap has been screenprinted using the commercial dot process. The use of intense colours lifts a rather banal image into the world of the bizarre.

Collotype

In 1855 Alphonse Poitevin patented the process of producing a printing plate from a surface covered in bichromated gelatine. A solution of light-sensitised gelatine is poured over a sheet of plate glass. When dry the plate is placed in contact with the negative and exposed to light. The dark parts in the original image (light in the negative) cause the gelatine to harden and become impervious to moisture, while the light parts in the original (dark in the negative) remain soft and absorbent. When the printing ink is applied to the gelatine it is accepted in inverse proportion to the amount of moisture the surface retains, the dryest areas accepting most and therefore printing darkest. The printing surface is formed of a massing of tiny irregular crinkles created as the gelatine dries in the initial plate-making process. The process has the advantage that it can render continuous gradations of tone without the use of a screen but has the disadvantage that the plate wears rapidly.

Crossline screen see half-tone.

Daguerreotype

Invented in 1839 by Louis Jacques Mandé Daguerre (1781–1851), who developed the ideas of J. N. Niépce. A silvered copper plate is sensitised by exposing it to iodine vapour which darkens the silver mirror surface, forming silver iodide. After exposure in a camera the plate is held over heated mercury vapour which forms an amalgam with the silver particles which have re-emerged from the iodide on its exposure to light. The excess silver iodide is washed away, leaving a surface slightly darker than the bright mirror-like quality of the mercury/silver amalgam. The resulting image is positive, very finely detailed and, after toning with gold salt, reasonably permanent provided the surface is protected. Multiplication of the unique daguerreotype image was attempted most successfully by Hippolyte Fizeau, who electrotyped and etched some examples in 1841. He was also the inventor of the gold toning process.

Diazo lithography

Modern alternative to transfer paper in lithography. A special aluminium lithographic plate is covered with a light sensitive coating. The artist draws on transparent film which is developed on this plate. There is no loss of autographic quality in this technique, which is considered by some to be more faithful than the traditional transfer paper method. It has the incidental advantage, exploited by Henry Moore for example, that the transparent sheets can be re-used or cut-up and rearranged as the basis of a new composition.

Dye Transfer

A method of producing a photographic print with greater colour density and more permanence than the normal emulsion process. The printing surfaces are formed by gelatine matrices, one for each colour separation. The matrices are immersed in dyes of the appropriate colours and printed in sequence, in exact register, on to specially prepared absorbent paper. The process has been used in fashion photography to alter the colours of an original image; this is virtually impossible by conventional photographic processing.

Electrotyping

Electro-chemical deposition of metal on to a surface in order to make a mould or replica of an object with extremely fine detail. Discovered in 1839, it was widely used for making copies of wood-engraved blocks. Many so-called wood-engravings are in fact printed from electrotypes.

Engraving

The action of incising lines with a graver into the surface of a plate to be used for printing. The gouged lines hold the ink. The term is often used loosely for many kinds of graphic work.

Etching

The action of chemically creating lines in the surface of a printing plate by selectively biting it with acid, as opposed to removing metal physically with a graver.

Gelatine

Transparent water soluble substance obtained from boiling animal skins and bones. Used as a medium to carry light sensitive materials and pigments. It is still essential in photography and photo-mechanical printing processes.

Half-tone

The process of photographically reproducing a tonal image by the use of a crossline screen interposed between the object to be photographed and the film. The range of tones in the original is interpreted on the negative by a mass of dots equally spaced but varying in size, giving the illusion of continuous tone. A half-tone block is a relief printing plate made from such a negative.

Ink photo

A combination of collotype and lithography developed for rapid reproduction of drawings and plans. A collotype is made of the drawing to give a fine tonal reproduction and is then transferred to a lithographic surface.

Intaglio printing

The process of printing from a plate in which ink is held in depressions in its surface. To take an impression, a dampened sheet of paper is laid on the inked plate and together they are submitted to sufficient pressure to drive the paper into the grooves so that it picks up the ink. Intaglio processes include line engraving and etching, mezzotint and aquatint, and photogravure.

Line block

Photographically produced relief printing surface perfected by the early 1880s. A zinc plate is coated with light sensitive gelatine and exposed to light through a negative. The gelatine is washed away where it is unhardened. The remaining sticky gelatine is dusted with asphaltum to protect it from the acid and the plate is etched in relief. This process can only be used for the reproduction of line drawings or flat areas with no intermediary tones: but if a high contrast negative is used a crisp black and white image, often wanted by artists, can be made. The effect of tonal gradation has been achieved by the later development of the half-tone letterpress block which combines the half-tone screen with the line block process.

Lithography

Lithography was invented by Senefelder in 1797. It depends on the mutual antipathy of grease and water. The drawing is made in a greasy substance on closely textured porous limestone or on zinc or aluminium plates. The printing surface is then moistened with the result that the greasy areas repel the water and the untouched parts absorb it. When greasy ink is applied, it is rejected by the wet areas and accepted by the greasy ones. Paper pressed on will, thus, pick up ink only from the area which has been drawn on.

Offset lithography

A system whereby the image on a lithographic stone or plate is printed on to an intermediate rubber roller, which subsequently deposits the image on to the paper. The image is therefore printed the same way round as it was drawn.

Mezzotint

A form of tonal engraving; the technique was widely used for the reproduction of oil paintings in the 18th and 19th centuries. The plate is prepared so that it will print an even deep black. This is done by systematically pitting its surface with a serrated chisel-like tool, known as a rocker, which raises a uniform burr. The design is formed by smoothing the burr so that different areas of the plate will hold different quantities of ink and therefore print different tones of grey.

Multiple

An edition of identical mass-produced art objects.

Negative

Certain silver salts darken with the action of light, so an image recorded photographically in silver has the tones in reverse (dark where it has been exposed to light and vice versa). By taking another image from this negative (as Sir John Herschel first termed it) the tones are returned to their correct value. Fox Talbot patented the Calotype process which used this principle in 1841.

Photogalvanography

A process that combines collotype with electrotype to produce an intaglio plate, patented by Paul Pretsch in 1854. A plate coated with light-sensitive bichromated gelatine is exposed through a negative. When wetted the gelatine swells most where it has been least exposed. This film of varying relief is electrotyped, and the resulting copy electrotyped again to form an intaglio plate. Used sporadically until the end of the 19th century, the process was finally replaced by photogravure.

Photogenic drawing

The process discovered by Fox Talbot in 1834 of making negative images by placing objects on sensitised paper and exposing them to light. Camera obscura images were also recorded in this way. His subsequent discovery of how to develop a latent image and to make positives led to his patenting the Calotype.

Photogravure (also known as Heligravure)

An intaglio printing process derived from Alphonse Poitevin's use in 1855 of carbon mixed with bichromated gelatine, which was further developed by Sir Joseph William Swan into his carbon print process. It is based on the discovery that light sensitised gelatine hardens and becomes insoluble in proportion to the amount of light that falls on it. This allows a gelatine tissue of variable thickness to be produced from a photograph. In hand photogravure the gelatine tissue is applied to a plate prepared with a fine aquatint ground and etched. As the acid bites deepest where the gelatine is thinnest, the plate is most deeply etched where least light has struck the original film, that is, where the darkest tones were in the original image. Machine photogravure, developed in the 1880s, utilizes the invention of the Czech Karl Klič (1841–1926) and others of a cross line screen printed on to the gelatine film before it is exposed under the negative, obviating the need for the aquatint ground. A flat-bed press can be used to print all forms of photogravure but machine photogravure is generally printed from a cylinder in a rotary press. Rotary gravure allows an enormous number of impressions to be printed at great speed. As the quality of the image is reasonably good even on cheap paper, the process is much used for magazines, catalogues etc.

Photo-lithography

In the 1830s J. N. Niépce experimented with sensitising a lithographic stone with bitumen mixed with oil of lavender. His method was improved by Barreswil, Davanne, Lerebours and Lemercier who published some photo-lithographs from photographs of architectural subjects in 1852–53. In 1855 Poitevin discovered that bichromated gelatine was a better coating for the stone. If a negative was exposed on the treated surface the unhardened gelatine could be washed away. The hardened area representing the design would act just like the greasy crayon, enabling reasonable reproductions of photographs to be made. In the 1880s the half-tone screen was introduced to give a more reliable range of tones.

Phototype

Photographically produced relief printing block, introduced by Messrs Fruwirth and Hawkins in 1868, which preceded the line block. A plate was prepared with a coating of bichromated gelatine and exposed under a negative. The unexposed and, therefore, unhardened gelatine was washed away, and the remainder electrotyped. The copper thus deposited over the surface, following the depressions formed by the lines of the image, created a relief block.

Posterisation

A method of transforming the subtle gradations in density of the tones of an original into several distinct steps of tonal density, without the use of a half-tone screen. At least three exposures of black and white film are used to obtain the various tones from light to dark within each of the customary three or four colour separations of commercial practice. Thus the technique requires a minimum twelve colour printings. The technique was pioneered in screenprinting at the Kelpra Studio.

Registration

The placing of successive blocks, plates or screens in the press so that they print the separate colours of an image accurately, one on top of the other. Sometimes, out-of-register effects are deliberately sought.

Relief printing

The process of printing from a surface raised above the areas to remain blank.

Screenprinting

Screenprinting is a stencil process in which the stencil is affixed to a fine mesh of silk, man-made fibre or steel, known as the 'screen'. The screen is stretched over an open frame. The ink is pushed across its surface by a flexible blade (squeegee), and forced through the holes in the mesh where it is not masked by the stencil on to the paper below. The stencil can be cut from paper or a transparent film specially prepared to adhere to the screen, or painted on in a fluid which makes the mesh impermeable. It can also be made photographically and be combined with the half-tone system, using light sensitised gelatine. Once the image has been developed on the light sensitised surface the soft parts of the gelatine, which correspond to where the ink is required to pass through the screen, wash away. Screenprinting was developed as a commercial printing process early in the 20th century but it was not used as a fine art medium until 1938 in America.

Separation

Preparation of negatives for full colour printing by photographic means. The coloured image is photographed through a series of filters which extract the areas to be printed by the respective primaries: cyan, magenta and yellow, to make separation negatives. In ink-printed images a fourth, black, negative is usually also required. From the negatives a separate printing plate or silk screen is made for each colour. When overprinted in succession an effect of full colour is obtained.

Solarisation

Over-exposure of a negative or positive photograph which in certain conditions results in the partial reversal of the tones. Some artists, Richard Hamilton for example, have exploited the effect for its own end.

Transfer paper
Paper with a specially prepared surface from which a lithographic
drawing can be transferred to a stone or plate.

Wood-engraving
Relief printing process from a block of end-grain wood worked with a
graver. First fully exploited at the end of the 18th century, it became
the most popular medium for commercial illustration until its place
was taken by photo-mechanical methods at the end of the 19th
century.

Woodburytype
An adaptation of the Carbon print process by W. W. Woodbury
(1834–1885) to provide mass-produced unfadeable alternatives to
albumen and collodion prints. Light sensitised gelatine is mixed with
carbon pigment and spread thickly on to a glass plate. Exposed
under strong light through a negative, it emerges as a film thickest in
the darkest areas and thinnest in the lightest. A mould is made by
pressing the gelatine into sheet lead under enormous pressure. Into
this mould is poured a mixture of water, pigment and ordinary
gelatine which dries to form the Woodburytype. Although at first sight
it looks like a photograph, it does not have any light sensitive material
in its final form. It is an image formed only of pigment suspended in
gelatine.

Record of the Exhibition

Measurements in centimetres, height before width

Anonymous
Architectural study, *c.*1929
Three stages in the half-tone relief block process
Original screened negative, etched plate, proof
23 × 15; 27.8 × 18, 28 × 17.5
Lent by the Science Museum

After Léon-Louis Chapon (c.1836–after 1900)
Girl with a pet bird. Bound up in
The Illustrated Photographer, August 14, 1868
Phototype, enlarged from a woodcut
26.6 × 21
VAM

After Albrech Dürer (1471–1528)
St John the Baptist and his executioner
Collotype
25.3 × 16.2
E.2668–1902

After A. Lavezzari (fl. *c.*1900)
Progressive stages in tri-colour printing process (6)
Screened photogravure
each: 11 × 7.8
E.2675–1680–1902

After Vincent Van Gogh (1853–1890)
Sorrow, 1885
Collotype
36.3 × 22.2
E.4159–1919

Ivor Abrahams (b. 1935)
Print from E. A. Poe, *Tales and Poems*, 1976
Four progressive proofs
Screenprints
18.3 × 23.5; 16.2 × 19.1; 16.2 × 18.9; 16.2 × 18.8
Lent by Christopher Betambeau

Norman Ackroyd (b.1938)
Plates (2) from *Landscapes & Figures*, eight etchings by Norman Ackroyd, eight
poems by William McIlvanney, 1973
Aquatint
16.5 × 13.8; 12.8 × 14.5
Lent by the Artist

Sir Lawrence Alma Tadema, O.M. R.A. (1836–1912)
Caracalla and Geta. After the painting dated 1907
Photogravure
74.5 × 87.7
E.174–1970

Barbara Astman (born 1950)
Carol with red blossoms . . . , 1975
Xerox
21.5 × 28
E.1170–1979

Carol in a plum living room . . . 1977
Xerox
14 × 20.5
E.1171–1979

Conrad Atkinson
Anniversary print from the people who brought you thalidomide . . ., 1978
Lithograph
53.4 × 44
E.1223–1979

Herbert Bayer (b.1900)
Lonely Metropolitan, 1932
Photomontage
38.7 × 28
Circ. 738–1968

Aubrey Vincent Beardsley (1872–1898)
The Kiss of Judas, 1893
Indian ink design, with an impression from the line block
31.1 × 29.9
E.292-1972

Joseph Beuys (b.1921)
How to explain paintings to a dead hare, 1970
Offset lithograph
18.4 × 11.6
Lent by the Arts Council

Peter Blake (b.1932)
The Beach Boys, 1964
Screenprint
53.5 × 32.2
Circ. 377-1965

Derek Boshier (b.1937)
Untitled. 1973
Etching
49 × 37.5
E.93-1974

Marcel Broodthaers (1924–1976)
Citron-Citroen, 1974
Offset lithograph and screenprint
97.5 × 61.3
E.1165-1979

La soupe de Daguerre, 1975
Colour photographs and screenprint
46.5 × 44.6
E.1132-1979

Nils Burwitz (b.1940)
Namibia: Heads or Tails? 1979
Screenprint
53 × 52
E.899-1979

Bob Chaplin (b.1947)
Pathway to Firle, 1976
Etching and aquatint
Two plates: 18.7 × 29.6; 21.4 × 29.8
E.418-1976

Seven Sisters. 1979
Screenprint and etching
44 × 71
E.1166-1979

Christo (b.1935)
Wrapped monument to Leonardo (2). 1970
Lithograph and photo-lithograph
Both: 74.5 × 55.5
Circ. 2 & 3-1972

A. F. J. Claudet (1797–1867)
Portrait of Mrs Pritchard, 1867
Hand-coloured daguerreotype
6.6 × 5.2
1422-1939

Harold Cohen (b.1928)
Richard III, 1967
Screenprint
65 × 73.5
Circ. 748-1969

Jean Baptiste Camille Corot (1796–1875)
La Petite Soeur. 1854
Cliché verre
15 × 18.7
E.2923-1921

Michael Craig-Martin (b.1941)
Untitled (2). 1970
Photo-lithograph
Both: 41.7 × 54.3
Circ.217 & 218-1970

Jennifer Dickson (b.1936)
Narcissus bound: Sleep, 1979
Xerox
21.5 × 28
E.1169-1979

Jim Dine (b.1935)
'Donald Duck' and 'Smoke in a vice'. Prints (2)
from the portfolio *Tool Box*, 1966
Screenprint and collage
Both: 60 × 47.3
Lent by Joe Studholme

Henry Dixon (1820–1893)
Oxford Arms, Warwick Lane, London, 1875
Albumen print
21.4 × 27.7
76-832

Oxford Arms, Warwick Lane, London, 1875
Carbon print
18.6 × 23.7
66-1958

Peter Henry Emerson (1856–1936)
The Sedge Harvest
Plate from the portfolio *Idyls of the Norfolk Broads*
Published by the Autotype Co, London, 1890
Autogravure
8.7 × 23

Max Ernst (b.1891)
Le Lion de Belfort. Plate from Une Semaine de Bonté; ou, les *Septs Elements*
Capitaux: Roman, Paris, 1934
Line block
18.7 × 14.7
Circ.156-1967

Frederick Evans (1852–1943)
Lincoln Cathedral: view from the castle, 1898
Photogravure
21 × 16
593-1903

Hippolyte Fizeau (1819–1896)
Portrait of two youths, *c.*1845
Impression from etched daguerreotype
14 × 19
Lent by the Royal Photographic Society

David Freed (b.1935)
Sounds, *c.*1975
Etching, aquatint and screenprint
45 × 89
E.80-1978

John Furnival (born 1933) & **John Vincé**
'Just for the record', 1978
Screenprint
52.8 × 25
E.1167-1979

Henry Gaudier-Brzeska (1891–1915)
Christmas card: a monkey, 1912
Zinc line block, with an impression from it
7.7 × 7.4
E.1970 & 1971-1946

Richard Hamilton (b.1922)
Adonis in Y fronts, 1962
Screenprint
68.6 × 82.5
Circ.59–1964

My Marilyn, 1965
Screenprint
51.5 × 63.1
Circ.130–1968

The critic laughs, 1968
Offset lithograph, laminated and retouched with enamel paint, and screenprint
59.5 × 46.5
Lent by the Artist

Bathers (b), 1969
Dye transfer
38.7 × 54.1
Lent by the Artist

Vignette, 1969
Dye transfer
48.5 × 39
Lent by the Artist

Kent State, 1970
Screenprint
67.3 × 87
Circ.212–1971

Portrait of the Artist by Francis Bacon, 1970–71
Screenprint over collotype
54.8 × 48.8
Circ.486–1976

John Heartfield (1891–1968)
Cover of *Jedermann sein eigner Fussball*
15 February 1919
Half-tone letterpress from photomontage
Lent by the British Library

Hurrah, die Butter is alle!
[Hurrah, the butter is finished!] 1935
Photomontage reproduced in *Arbeiter-Illustrierte-Zeitung*
Collotype
55 × 37
C.25611

William Harcourt Hooper (1834–1912)
Woodblocks with designs from Holbein's 'Dance of Death'
photographed on to them
Death seizing a King: photo-sensitised surface uncut
Death seizing the Pope: woodblock partially cut
Death and the labourer: woodblock fully cut
Each 7 × 5.8 × 2.2
E.1295, 1294, 1292–1912

Horne & Thornthwaite, London
Portrait of an old man.c.1861
Hand-coloured carte-de-visite photograph
9.1 × 5.8
VAM

Alyson Hunter (b.1948)
Window. '71', 1971–74
Photogravure
40.3 × 24.2
E.2213–1974

Angela Jansen
Bird Cage, 1973
Etching and photogravure
30.9 × 14.9
E.12–1977

Jasper Johns (b.1930)
Four panels from Untitled 1972, 1973–74
Lithograph embossed by photo-etched plates
103.5 × 73, 104 × 73, 102.5 × 73, 102.5 × 73
Circ. 355–8–1976

Frontispiece for *Fizzles* by Samuel Beckett, 1976
Colour lithograph
Published by Petersburg Press, London 1976
L.1623–1977

Allen Jones (b.1937)
Secret Valentines, 1962
Lithograph
36.3 × 59.9
Circ.308–1963

Prints (2) from 'Life Class' suite, 1968
Lithographs
34.9 × 56.3; 46.5 × 56.3
Circ.136 & A–1973

E. McKnight Kauffer (1890–1954)
Poster advertising Miles-Whitney Straight Aircraft and Shell Lubricating oil, 1937
Offset lithograph
73.3 × 111
E.577–1971

Michael Kidner (b.1917)
Cross Stretch One, 1977
Etching
34.2 × 44.2
E.82–1979

Ronald B. Kitaj (b.1932)
Acheson Go Home, 1963
Screenprint, with original collage
73.1 × 53.1; 38.3 × 27.6
Circ.198–1964 & 183–1964

The Cultural Value of Fear, Distrust and Hypochondria. From the series 'Mahler
becomes Politics, Beisbol', 1966
Screenprint
52.3 × 77.1
Circ.401–1966

'Transition', 'Towards a Better Life' and 'Photo-Eye'. Nos. 5, 18 and 21 from the
series 'In Our Time; covers for a small library after the life for the most part', 1969
Screenprints
77.6 × 55.6; 45.2 × 37.5; 53.8 × 38
Circ. 189, 187 & 217–1975

Gerald Laing (b.1936)
A.A.D., 1968
Screenprint
50.2 × 76.8
Circ. 667–1968

Lemercier, Lerebours, Barreswil & Davanne
St Loup de Naud. 1852—3. After a photograph by Lescq, dated 1851
Lithograph
33 × 23
E.1182 B–1887

Gail McKennis (b.1939)
Janet and the Vermeer II. From the series 'Art of the Real', 1972
Lithograph
50.8 × 41.3
E.508–1972

Mayall & Co, London
Portrait of H.R.H. Princess of Wales, c.1861
Hand-coloured carte-de-visite photograph
9.1 × 5.9
3546–1953

J. Meisenbach Co Ltd, London (active 1884–c.1914)
'Rival Belles', c.1895. Advertising sheet for process engraving
22.7 × 14
E.9736–1905

Edward Meneely (b.1927)
A portrait of Jasper Johns, 1967
Electrostatic print
36.8 × 22.8
Circ. 204A–1971

Lazlo Moholy-Nagy (1895–1946)
The Shooting Gallery, 1925. (Print, 1973)
Photomontage
27.4 × 37
Circ. 205–1974

Henry Moore (b.1898)
Woman holding cat, 1949
Collograph
29.8 × 48.9
Lent by the Artist

Two women bathing a child I, 1973
Lithograph, with Diazo transparencies
54.3 × 60.2
E.1408–1976

Malcolm Morley (b.1931)
Beach Scene, 1969
Screenprint
42 × 54.8
Circ.18–1969

Jim Nawara (b.1945)
Horseshoe mound, 1974
Cliché verre
35.5 × 45.7
Lent by Cartwright Hall, Bradford Art Galleries and Museums

Joseph Nicéphore Niépce (1765–1843)
Portrait of the Cardinal d'Amboise, 1827
Heliogravure plate, with an impression from it
18 × 13.9
Lent by the Royal Photographic Society

Eduardo Paolozzi (b.1924)
Metalisation of a dream, 1963
Screenprint
50.6 × 48.7
Circ. 155–1964

Hollywood Wax Museum. From the set 'Zero Energy Experimental Pile', 1969–70
Screenprint on Astralux, laminated in clear acrylic sheet
77 × 47
E.1700–1977

Carpenter. From 'The Conditional Probability Machine', 1970
Photogravure
19.8 × 12.5
Lent by the Artist

Chimpanzees in a Test Box Designed for space flight and Mobot Mk I from 'Cloud Atomic Laboratory', 1971. With source material
Photogravure
Circ. 673–1971

Nettleton. From 'Calcium Light Night'. 1974–76
Screenprint, with original collage and film
99 × 69
Lent by Christopher Betambeau & E.1156–1978

Victor Pasmore (b.1908)
When the lute is broken . . . , 1974
Etching and aquatint
38.2 × 39.3
E.2172–1974

Peter Phillips (b.1939)
The Lion. From the portfolio 'Pneumatics', 1968
Screenprint
59.5 × 94.6
Circ. 105–1969

Tom Phillips (b.1937)
'Sixty-four stopcock box lids' and 'Matching colours struck by heatwave'. From the set 'The walk to the studio'. 1976
Screenprints
Both: 101.3 × 71.1
E.998–1979; Lent by Waddington Galleries Ltd

John Piper (b.1903)
Chapel of St George, Kemptown, 1939
Etching and aquatint
19.6 × 27.7
E.2092–1974

Llagloffan, Pembroke: Baptist Chapel from
A Retrospect of Churches, 1964
Lithograph
59.4 × 81.9
Lent by Stanley Jones

Chris Plowman (b. 1952)
Electric Storm, 1976
Etching and airbrushed watercolour
37.4 × 26.7
E.654–1976

Alphonse Poitevin (1819–1882)
Landscape in southern France, 1855. After a photograph by F. Teynard
Lithograph
22 × 30.5

C. R. Pottinger, Cheltenham
Portrait of a young man in fancy dress, c.1861
Hand-coloured carte-de-visite photograph
9.4 × 5.8
678–1956

Paul Pretsch (1808–1873)
Don Quixote in his study, after a photograph by
William Lake Price. From *Photographic Art Treasures*, London 1857
Photo-Galvano-Graphic process engraving
22.6 × 19.7
34–299

William Lake Price (1810–1896)
Don Quixote in his study, 1855
Albumen print
32.4 × 28.3
3–1976

Robert Rauschenberg (b.1925)
License, 1962
Lithograph
100 × 71
Circ. 120–1964

Man Ray (1890–1976)
Untitled Rayograph, 1921–28
Photogram
29 × 21.5
Circ. 708–1964

Michael Rothenstein (b.1908)
Clench, 1969
Screen and relief print
81.3 × 58.4
Lent by the Artist

Colin Self (b.1941)
Power and Beauty Nos. 1 and 3, 1968
Etchings
69.5 × 105.7; 68 × 104
Circ. 96 & 98–1969

Richard Smith (b.1931)
Zoom, 1964
Screenprint
48.2 × 76.3
Circ. 169–1964

Butterfly 6, 1972
Etching
63.2 × 91
Lent by the Arts Council

Antoni Starczewski (b.1929)
MF 30/g, 1973
Blind relief etching
39.7 × 39.7
E.1549–1977

Norman Stevens (b.1937)
Path, 1974
Etching
32.6 × 62.5
Lent by the Artist

William Henry Fox Talbot (1800–1877)
Lace, 1842
Calotype, negative and positive
Both: 18.4 × 22.2
Lent by the Science Museum

Plant study, *c.*1850
Photoglyphic engraving plate, screened with muslin and proof
Both: 10.2 × 6.3
Lent by the Science Museum

Portal of St Trophimus, Arles, 1866
Photoglyphic engraved plate and proof
20.4 × 25.5
Lent by the Science Museum

Felix Teynard (Active 1850–1860)
Landscape in southern France, 1851
Albumen print
22.7 × 30.5
VAM

John Thompson (1837–1921)
Street Advertising. Illustration from *Street Life in London*, published by Samson Low, Searle and Rivington, London 1877
Woodburytype
11.1 × 8.7
VAM

Norman Thompson (b.1951)
A definition of a print, 1974
Etching and screenprint
48.5 × 47.5
Lent by Cartwright Hall, Bradford Art Galleries and Museums

Joe Tilson (b.1928)
Transparency Clip-O-Matic Lips, 1967
Screenprint
71 × 50.8
Lent by the Arts Council

Five objects in space, 1966
Screenprint on acrylic sheet
58 × 88.6
Circ. 776–1966

William Tillyer (b.1938)
The Large Bird House, 1971
Etching
47.5 × 35.7
E.579–1971

David Tremlett (b.1941)
Brickwork, 1970
Screenprint
102.2 × 68.9
Circ. 197–1970

Ger Van Elk (b.1945)
The Symmetry of Diplomacy, The Stanhope, New York, 1975
Photograph and screenprint
22.6× 41
VAM

John Walker (b.1939)
Juggernaut, 1974
Screenprint
86.6 × 145
E.429–1975

Andy Warhol (b.1930)
Marilyn Monroe. Prints (2) from 'Ten Marilyns', 1967
Screenprints
Each: 91.4 × 91.6
Circ. 121, 123–1968

Jackie Four Times, 1966
Screenprint
101.6 × 76.3
Circ. 591–1968

Donald Wilkinson (b.1937)
The valley behind me. From 'Mr Gray's Journal', 1974
Etching and aquatint
58.5 × 49.2
E.1730–1974

Gerd Winner (b.1936)
New York Canyon, 1974
Screenprint
98.7 × 61.5
E.83–1974